Lecture Notes in Mathematics

A collection of informal reports and seminars
Edited by A. Dold, Heidelberg and B. Eckmann, Zürich

316

Symposium on Non-Well-Posed Problems and Logarithmic Convexity

Held in Heriot-Watt University, Edinburgh/Scotland
March 22–24, 1972

Edited by R. J. Knops
Heriot-Watt University, Edinburgh/Scotland

Springer-Verlag
Berlin · Heidelberg · New York 1973

AMS Subject Classifications (1970): 35R25, 35B30

ISBN 3-540-06159-2 Springer-Verlag Berlin · Heidelberg · New York
ISBN 0-387-06159-2 Springer-Verlag New York · Heidelberg · Berlin

Offsetdruck: Julius Beltz, Hemsbach/Bergstr.

PREFACE

The following articles represent the invited lectures given at the Symposium on Non-Well-Posed Problems and Logarithmic Convexity held at Heriot-Watt University, Edinburgh, Scotland from March 22-24, 1972. The Symposium was organised jointly by the University of Dundee and Heriot-Watt University, and formed part of the activities of the North British Differential Equations Symposium for the academic year 1971/72. These activities were sponsored by the Science Research Council of Great Britain.

One of the main objects of the Heriot-Watt Symposium was to provide expository accounts of recent developments in the subject which would be of use not only to the specialist but also to someone wishing to become acquainted with this area of activity in partial differential equations. The following collection of addresses therefore contains both surveys and discussions of current problems and it is hoped they will further serve to fulfill the objects of the Symposium.

The organisers take this opportunity of thanking Heriot-Watt University for the invitation that made the Symposium possible, and also for providing generous hospitality. They would also like to thank the S.R.C. of Great Britain. Finally, they gratefully acknowledge the advice and assistance given by Professor L.E. Payne in the preparation of the programme, and wish to express their appreciation to Miss S. Corbey for her help in the preparation and typing of the manuscripts.

R.J. Knops
Editor

CONTENTS

Lectures whose proceedings do not appear here

R. Khosrovshahi: Growth properties of solutions of Schrodinger type systems

SOME GENERAL REMARKS ON IMPROPERLY POSED PROBLEMS FOR PARTIAL DIFFERENTIAL EQUATIONS

L.E. PAYNE

INTRODUCTION

The aim of this paper is to present some historical remarks and a not-too-technical introduction to the subject of improperly posed problems for partial differential equations. We shall indicate various methods which have been employed for handling such problems - methods whose power and applicability are discussed in the accompanying papers in this volume.

A rather rough definition of a well posed problem for differential equations is as follows: A problem is said to be *well posed* (or *properly posed*) if a unique solution exists which depends continuously on the data. Of course, we must state precisely in what class the solution is to lie as well as the measure of continuous dependence. A problem which is not well posed is said to be *ill posed*, *non-well posed*, or *improperly posed*. Some simple improperly posed problems are the final value problem for the heat equation, the Cauchy problem for the Laplace equation, and the Dirichlet problem for the wave equation. It should perhaps be emphasised at this point that in discussing the question of continuous dependence on the data we must consider as "data" any initial or boundary values, prescribed values of the operator, coefficients of the equation as well as the geometry of the domain of definition.

It is difficult to say when mathematicians first became concerned over the question of proper posedness of boundary and initial value problems. Most physically reasonable problems seem to lead to well posed problems for the corresponding mathematical models. Equilibrium problems lead to elliptic equations with boundary data required in both the physical and mathematical contexts. Wave or vibration problems generally lead to hyperbolic equations, and initial or initial-boundary value problems are suggested by the physics. Diffusion or evolutionary processes are generally described by parabolic equations, and it is again initial data that the physicist requires. Until recently, the prevailing attitude toward the study of classes of improperly posed problems, which may be found expressed, for instance, in Petrowsky's book [121], was that fortunately there is no need to consider such problems since they are of no physical interest. Unfortunately, as we are now well aware, there are many physical situations which force us (if we are to get any answers at all) to consider problems that are improperly posed.

Although it is clear from the work of Cauchy, Kowaleski, Holmgren and others that more than a century ago some consideration was being given to the question of proper posedness, the father of this study is generally considered to be Hadamard [50]. It was he who, at the turn of the century, clearly defined the question and illustrated by examples and counter examples the difficulties involved. Among other things, he derived necessary and sufficient conditions for global existence of solutions of the Cauchy problem for the Laplace equation. Unfortunately, it is impossible to verify, in general, whether the necessary conditions are or are not satisfied.

Even today the question of global existence of solutions of the various classes of improperly posed problems is largely unsettled. From a practical point of view, one probably requires an existence theorem which is quite different from the standard type of theorem required for properly posed problems. Roughly speaking, what one really needs is a theorem which asserts the existence of a solution whose data are close in some measure to the given data and which in addition satisfies some appropriate auxiliary stabilising condition. Little has been done on the question of existence, the major effort being directed rather toward the simpler question of uniqueness, where prominent among the earlier work is the fundamental theorem of Holmgren [53] and some results due to Carleman [23] in the late thirties. However, a large number of uniqueness proofs for improperly posed Cauchy problems began to appear in the literature of the 1950's and early 1960's, frequently as consequences of unique continuation theorems. Among the numerous mathematicians who have contributed in this area are Agmon [3,4], Agmon and Nirenberg [5], Aronszajn [7], Calderon [16], Cordes [28], Douglis [34,35], Foias, Gussi and Poenaru [43], Heinz [51], Hörmander [54], John [60], Kumano-go [75], Landis [76], Lavrentiev [79,80], Lions and Malgrange [89], Lopatinskii [90], Mizohata [102,103], Miller [104], Nirenberg [107], Pederson [119], Plis [123], Protter [125], Pucci [129,130], Shirota [137] and many others (see e.g. Lattes and Lions [77], Lavrentiev [80,81], Payne [113] and the bibliographies contained in these references).

These authors have dealt almost exclusively with the question of uniqueness. Far more important, of course, are the questions of continuous dependence (which implies uniqueness) and approximation of solutions. In this paper, we first make brief mention of some different classes of improperly posed problems, make a few brief remarks on the question of existence and then confine our discussion to classes of improperly posed Cauchy problems. For such problems we mention various methods which have been used to establish uniqueness and continuous dependence, and also to determine approximate solutions for the problems in question. Finally, we demonstrate, using a simple example, the principal features of several of these methods.

Mathematicians recognised very early that many problems of physical interest were completely intractable when tackled directly. They frequently adopted an inverse or semi-inverse method of trying to generate a reasonable approximate solution of the problem in question by combining simpler solutions of simpler problems. The semi-inverse method was used very successfully by St. Venant in his formulation of classical torsion and flexure problems, and fluid dynamicists through the years have tried to generate flows past obstacles by combining in an ingenious way the flows generated by various distributions of singularities. In the late 1930's mathematicians began to concentrate on the many mathematical questions raised in the treatment of these various types of inverse problems. Since that time much work has been done, particularly by the Russians, starting with Novikov [108], Rapaport [133], Sretenskii [138], Tihonov [139] and others and continuing with Lavrentiev [80], Lavrentiev, Romanov and Vasiliev [81], Berzanskii [11], Marčuk [92,93] and many others. (Reference [81] provides a recent account of the Russian literature on this subject.)

One of the most interesting classes of inverse problems is that in which coefficients are to be determined from a knowledge of certain functionals of the solution. In general, such problems fall into the category of improperly posed problems.

As an example of this latter type of problem, suppose we are given an initial boundary value problem for the equation

$$a \frac{\partial u}{\partial t} = \Delta u$$

where Δ is the Laplace operator and "a" is an unknown constant. If, in addition to appropriate initial and boundary data, the value of the solution at other suitable points in space-time is prescribed, will it be possible to determine the coefficient "a"? If so, is it possible to compute necessary and sufficient criteria? Such problems have been studied extensively in Russia but they have also attracted the attention of such authors as Cahen [15], Cannon [17,19,20], Cannon and Dunninger [21], Douglas and Jones [33], and Jones [63]. (See also e.g. Lavrentiev, Romanov and Vasiliev [81], Lattes and Lions [77], and the papers cited therein.) The work mentioned above deals mainly with stabilising conditions and procedures for determining or approximating the unknown coefficients.

Although we shall not be here concerned with boundary value problems for hyperbolic equations, it should at least be mentioned that until the past year or so very little work in this area has appeared in the literature. There was the early paper of Bourgin and Duffin [12] and subsequent papers by Abdul-Latif [1], Dunninger and Zachmonoglou [36], Fox and Pucci [44], John [58] and others, the investigations dealing primarily with questions of existence and uniqueness of classical solutions. A number of papers dealing with this class of problems are just now appearing in the literature and a summary of work in this area would seem appropriate at some later date.

At this point let us briefly mention how other types of improperly posed problems might arise. It often happens that we do not know precisely the region of definition of the equation governing our problem (an elastic solid with internal cracks, inhomogeneities or holes, whose locations are not precisely known). It also frequently happens that a portion of the boundary of our region is inaccessible for measuring the desired data, as in the problem of locating oil deposits below the surface of the earth (see [131]). It may also be the case that we need to locate certain internal singularities which we know to exist but whose precise position is unknown or that we wish to place certain singularities in positions which will guarantee a desired behaviour. In this latter category we might mention the problem of the electron focusing gun (see e.g. Hyman [55], Radley [55a]). One usually tries to deal with the above mentioned types of problems by measuring an over-abundance of data and then trying to compensate for the lack of knowledge of the geometry or location of singularities. However, extreme care must be employed since the resulting problems are almost invariably improperly posed. For additional physically important examples see [80] and [81].

As we mentioned earlier the existence question is an extremely difficult one, and we shall have very little to say about it. Before leaving the question we should, however, re-emphasise an observation of Fichera [42] that the frequently held assumption that a partial differential equation (even a linear one) with smooth coefficients possesses an infinite number of solutions is false. In fact, it may fail to have any non-trivial solutions at all. Fichera [42] discusses the example of Lewy [88] and an earlier simple example due to Picone [122]. We mention also the example of Plis [124].

Our subsequent remarks will in general be confined to improperly posed Cauchy problems or initial-boundary value problems. Here it is known for the typical problem that the solution will not exist globally unless strong compatibility relations hold among the data. Even if for the particular data of the problem the solution should exist, it will not depend continuously on the data in the sense of Hadamard [50]. We note, for instance, that Hadamard's condition for the existence of solution of the Cauchy problem for the Laplace equation was that a certain combination of the data be continuable as an analytic function throughout the domain of definition of the solution. Extensions of Hadamard's method have been made by Aziz, Gilbert and Howard [8], Colton [26], Garabedian and Lieberstein [48], Ivanov [56], Payne and Sather [118], D. Sather and J. Sather [134] and others. For additional discussion see the paper of Colton [25] in this volume as well as the books of Bergman [9], Gilbert [49] and Vekna [143].

In practise, however, the type of information necessary for determining the continuability is usually not available. The data are obtained by measurement (the geometry, coefficients, boundary values of the solution, etc.), and hence are not known precisely. Thus one is forced to work with data which is not precise and the results must reflect and allow for this possible error in data measurement. As mentioned earlier, Cauchy problems of this imprecise data type have been presented to mathematicians by petroleum engineers, meteorologists, physicists and scientists working in a wide variety of fields. (See for example the references cited in [80], [81] and [131].) In such problems it is frequently the case that the physical problem which one really would like to solve is a standard well posed problem, but a portion of the boundary is either unknown, extremely irregular or inaccessible for measuring the desired data. As we indicated before, one hopes that by measuring an overabundance of data on an accessible portion of the boundary he can compensate somehow for the lack of data elsewhere.

In [62] John pointed out that it is typical of such improperly posed problems that there exists at most one solution. We would therefore expect that by suitably restricting the class of solutions to be considered it should be possible to bring about continuous data dependence. As John emphasises, the difference between properly and improperly posed problems is that the required restrictions for improperly posed problems cannot be inferred from the data alone or at least not from approximate knowledge of the data.

In the middle 1950's, John [59-62] and Pucci [128-132] published a series of papers deriving continuous dependence results in suitably restricted classes. In particular, they obtained stability results for various problems in the class of uniformly bounded solutions. Work along the same lines was continued by Douglas and Gallie [32], Lavrentiev [78-81], and in the 1960's by a host of authors (see e.g. the papers cited in Payne [113]). It should be noted that in a number of physically interesting problems this uniform bound (or appropriate norm bound) can be obtained by observation. Since it need not be sharp, any crude bound will suffice.

It is perhaps fitting to recall at this point that this conference is devoted to the two topics - improperly posed problems and convexity methods. Although the tool of logarithmic convexity has had many important applications in the study of improperly posed problems (see e.g. Agmon [4], Conlan and Trytten [27], Crooke [29], Edelstein [39], Hills [52], Knops and Payne [67-72], Knops and Steele [73], Lavrentiev [78-79], Levine [83-86], Miller [94], Payne [112-114], Payne and Sather [115-116], Schaefer [135-136], Trytten [142] and others) it has also been used in the study of questions such as the growth properties of solutions as some variable tends to infinity (see Agmon [4,5], Edmunds and Dyer [38,40], Khosrovshahi [64], Ogawa [110-111], and the references cited in the papers of Dyer [37] and Khosrovshahi [65] which appear in this volume). We should also mention that convexity arguments have been used by Adelson [2] and by Payne and Sather [112] to study certain "singular perturbation" questions for classes of improperly posed problems. At the same time it must be remarked that in addition to logarithmic convexity methods, many other methods have been used in the study of various types of improperly posed problems. The accompanying is a partial list of tools used to tackle such problems.

1. Function theoretic methods ([8], [26], [48], [49], [56], [143])
2. Eigenfunction methods ([17], [18], [95], [101])
3. Logarithmic convexity methods ([4], [27], [29], [39], [52], [67-72], [73], [83-86], [112-114], [115-116], [135], [136], [142])
4. Methods of Protter ([125], [126], [127], [105], [106])
5. Lagrange identity methods ([13], [14], [106a])

6. Quasireversibility methods ([46], [77])

7. Restriction of data to the class of band limit functions ([100], [144])

8. Numerical and programming methods ([31-33], [48], [59], [61], [93], [96-98])

This is not intended to be an exhaustive list but rather an indication of the variety of tools that have been used.

It is perhaps informative in an introductory paper such as this, to illustrate how several of these methods are applied, using as a model the initial-boundary value problem for the backward heat equation. By restricting attention to this simple model we will clearly be unable to indicate the power and wide applicability of the various methods. We can merely give some idea of the main arguments, and indicate the type of information on uniqueness, continuous dependence and growth of solution that can be obtained. More will be said about several of these methods in the other papers of this volume.

2. EIGENFUNCTION METHODS

The problem we shall consider is the following: let D be a simply connected closed bounded region in n-space with smooth boundary ∂D. We seek in D a C^2 solution $u(x,t)$ of the following problem

$$\begin{aligned} Lu \equiv \frac{\partial u}{\partial t} + \Delta u &= 0 \text{ in } D\ (0,T) \\ u &= 0 \text{ on } \partial D x [0,T] \qquad (1) \\ u(x,0) &= f(x) . \end{aligned}$$

We know that, in general, no solution exists and that if one does exist it will not, in general, depend continuously on the data. We know too that the stability problem can be overcome by suitably restricting the class of admissible data or the class of solutions to be considered.

An eigenfunction method involves expansion of the solution in terms of the eigenfunctions of the Laplace operator, and it is easily seen that if u_i

and λ_i are the ith normalised eigenfunction and the corresponding eigenvalue of

$$\begin{aligned} \Delta u + \lambda u &= 0 \text{ in } D \\ u &= 0 \text{ on } \partial D , \end{aligned} \tag{2}$$

then the formal solution of the problem is given by

$$u(x,t) \approx \sum_{n=1}^{\infty} f_n e^{\lambda_n t} u_n(x) \tag{3}$$

where f_n is the nth Fourier coefficient of f(x). Obviously, for this series to represent even an L_2 solution of (1) in Dx(0,T) it is necessary that the Fourier coefficients decay faster than $e^{-\lambda_n T}$ as $n \to \infty$. This is, of course, an extremely restrictive condition to impose on the initial data, particularly in view of the fact that f(x) was probably determined by measurement and hence subject to error.

A restriction of the data to band-limited functions is essentially equivalent to replacing f(x) by $f_n(x)$, the first m terms of its Fourier series expansion. The corresponding solution $\hat{u}(x,t)$ is then given by

$$\hat{u}(x,t) = \sum_{n=1}^{m} f_n e^{\lambda_n t} u_n(x) , \tag{4}$$

a perfectly well defined solution for the approximate data.

Now instead of replacing f(x) by $f_n(x)$, let us assume that we have made an error in measuring the initial data, calculating it to be $f^*(x)$ instead of the true value f(x). Suppose further that we can compute a constant K such that at some time T

$$\int_D u^2(x,T)dx \leq K^2 . \tag{5}$$

This constant K can frequently be determined from the physics described by the problem. Let

$$\int_D [f(x)-f^*(x)]^2 dx \leq \alpha^2 . \tag{6}$$

Thus our formal solution (3) now satisfies

$$\sum_{n=1}^{\infty} f_n^2 e^{2\lambda_n T} \leq K^2 \tag{7}$$

and

$$\sum_{n=1}^{\infty} \left[f_n - f_n^*\right]^2 \leq \alpha^2 . \tag{8}$$

Assuming K and α to be prescribed then any f(x) with Fourier coefficients satisfying (7) and (8) could be regarded as giving an acceptable "solution" of our problem. Clearly such a solution need not exist for arbitrary K and α and if one does exist it will not, in general, be unique.

3. QUASIREVERSIBILITY METHODS

The idea of quasireversibility is as follows: one alters the operator to make the problem a properly posed one, and using the solution of the altered problem as a guide one then constructs an approximate "solution" of the ill posed problem.

For instance let $\tilde{u}$ be a classical solution of

$$\begin{aligned} &\frac{\partial \tilde{u}}{\partial t} + \Delta\tilde{u} + \varepsilon^2\Delta^2 u = 0 \text{ in } Dx(0,T) \\ &\tilde{u} = 0,\ \Delta\tilde{u} = 0 \text{ on } \partial Dx\left[0,T\right] \\ &\tilde{u}(x,0) = f(x) . \end{aligned} \tag{9}$$

Under appropriate smoothness hypotheses the solution of this problem is known to exist and be unique. Formally the solution is expressible as

$$\tilde{u}(x,t) = \sum_{n=1}^{\infty} f_n e^{-\lambda_n(\varepsilon^2\lambda_n - 1)t} u_n(x) . \tag{10}$$

If for fixed ε^2, $f \varepsilon L_2$ then $\tilde{u} \varepsilon L_2$. The problem (9) is well posed and for sufficiently small ε the first several terms in the series (10) are essentially the same as the corresponding terms in (3).

At t = T we find

$$\tilde{u}(x,t) = \sum_{n=1}^{\infty} f_n e^{-\lambda_n(\varepsilon^2\lambda_n - 1)T} u_n(x) . \tag{11}$$

Let us now define a new problem, i.e. we seek a solution U(x,t) of

$$\begin{aligned}
&\frac{\partial U}{\partial t} + \Delta U = 0 \text{ in } Dx(0,T) \\
&U = 0 \text{ on } Dx[0,T] \\
&U(x,T) = \sum_{n=1}^{\infty} f_n e^{-\lambda_n(\varepsilon^2\lambda_n - 1)T} u_n(x) \equiv \tilde{u}(x,T) .
\end{aligned} \tag{12}$$

Problem (12) is properly set and the solution is given by

$$U(x,t) = \sum_{n=1}^{\infty} f_n e^{[-\lambda_n^2\varepsilon^2 T + \lambda_n t]} u_n(x) . \tag{13}$$

We wish now to see how close U(x,0) is to the initial data f(x). Thus

$$||U(x,0) - f(x)||^2 \equiv \int_D [U(x,0) - f(x)]^2 dx = \sum_{n=1}^{\infty} f_n^2 [1 - e^{-\lambda_n^2\varepsilon^2 T}]^2 . \tag{14}$$

In the quasireversibility method the function U(x,t) is used as the "solution" of the problem (3). Note that as $\varepsilon \to 0$ the right hand side of (14) tends to zero, but the limit of the right hand side of (11) may not exist. There is clearly much non-uniqueness in this procedure. In fact, it has been pointed out by Gajewski and Zacharias [46] that there are certain advantages to using instead of $\tilde{u}$, the solution $\hat{u}$ of the well posed problem

$$\begin{aligned}
&\frac{\partial}{\partial t} [\hat{u} - \varepsilon^2\Delta\hat{u}] + \Delta\hat{u} = 0 \text{ in } Dx(0,T) \\
&\hat{u} = 0 \text{ on } \partial Dx[0,T] \\
&\hat{u}(x,0) = f(x) .
\end{aligned} \tag{15}$$

The solution is then given by

$$u(x,t) = \sum_{n=1}^{\infty} f_n e^{\frac{\lambda_n}{1+\varepsilon^2\lambda_n} t} u_n(x) . \tag{16}$$

As before we now define our approximating solution $V(x,t)$ to be a solution of

$$\frac{\partial V}{\partial t} + \Delta V = 0 \text{ in } Dx(0,T)$$

$$V = 0 \text{ on } \partial Dx[0,T] \tag{17}$$

$$V(x,T) = \sum_{n=1}^{\infty} f_n e^{\frac{\lambda_n}{1+\varepsilon^2\lambda_n} T} u_n(x) .$$

The solution of (17) may be expressed as

$$V(x,T) = \sum_{n=1}^{\infty} f_n e^{-\left[\frac{\varepsilon^2\lambda_n^2 T}{1+\varepsilon^2\lambda_n} - \lambda_n t\right]} u_n(x) . \tag{18}$$

Thus in this case

$$||V(x,0) - f(x)|| = \sum_{n=1}^{\infty} f_n^2 \left[1-e^{\frac{\varepsilon^2\lambda_n^2 T}{1+\varepsilon^2\lambda_n}}\right]^2 . \tag{19}$$

Clearly the right hand side of (19) is smaller than the right hand side of (14) for any positive ε.

This is a slight indication of the need for guidelines in the application of the quasireversibility method. It should be emphasised of course that for a more complicated operator we might not be able to exhibit $\tilde{u}(x,t)$ and $U(x,t)$ explicitly in which case a bound for the left hand side of (14) might be more difficult to obtain. This method is discussed in fuller generality in the papers of Miller [99] which appear in this volume.

4. LOGARITHMIC CONVEXITY METHODS

The method of logarithmic convexity proceeds as follows: we seek a function $F(t)$ defined on solutions of (3) which satisfies the following properties:-

(i) $F(t) \geq 0$ for $0 \leq t < T$

(ii) $F(t) = 0 \Longleftrightarrow u(x,t) = 0$ for $0 \leq t < T$

(iii) $FF'' - (F')^2 \geq 0$ for $0 < t < T$.

Here the prime denotes differentiation with respect to t. It can be easily shown that if F vanishes at any point $t_1 \varepsilon [0,T]$ then $F(t) \equiv 0$ in $[0,T]$. Thus without loss one assumes that $F(t) > 0$ in $[0,T]$. Condition (iii) will then yield the two inequalities

$$F(t) \leq [F(T)]^{t/T}[F(0)]^{1-t/T} \tag{20}$$

and

$$F(t) \geq F(0)\exp(F'(0)/F(0)) \; . \tag{21}$$

Thus if one had the additional information that

$$F(T) \leq K^2 \tag{22}$$

for some prescribed K, (14) would lead to the Hölder stability result

$$F(t) \leq K^{2t/T}[F(0)]^{1-t/T} \qquad 0 \leq t \leq t_1 < T \; , \tag{23}$$

while (21) would give a lower bound for the growth (or decay) rate.

For our example (3) we choose

$$F(t) = (u,u) \equiv \int_D u^2 dx \tag{24}$$

and compute

$$\begin{aligned} F'(t) &= 2(u,u_t) \\ F''(t) &= 4(u_t,u_t) \; . \end{aligned} \tag{25}$$

Clearly then condition (iii) is satisfied (this follows from a simple application of Schwarz's inequality). Thus assuming knowledge of the K in (5) we see that (23) reduces to

$$\int_D u^2(x,t)dx \leq K^{\frac{2t}{T}} \left[\int_D f^2 dx\right]^{1-t/T} \; , \tag{26}$$

while (21) leads to the result

$$\int_D u^2(x,t)dx \geq \left[\int_D f^2 dx\right] \exp\left[\frac{2\int_D |\text{grad} f|^2 dx}{\int_D f^2 dx}\right] \; . \tag{27}$$

Since (26) represents an a priori upper bound for the L_2 integral of u it is not surprising that this method may be used to obtain explicit L_2 and pointwise bounds for the solution itself. For more details see the papers of Payne [112], Schaefer [135-136], Trytten [142], etc.

It is interesting that in this special example if one chooses

$$F_k(t) = (u,u_k)^2 \tag{28}$$

then

$$F_k F_k'' - [F_k']^2 \equiv 0 \tag{29}$$

and one obtains in fact the two identities

$$\begin{aligned} (u(x,t),u_k(x))^2 &= [f_k]^{2(1-t/T)}(u(x,T),u_k(x))^{2t/T} \\ (u(x,t),u_k(x))^2 &= f_k^2 e^{2\lambda_k t} . \end{aligned} \tag{30}$$

It should be noted that the inequality (29) holds in this case because of the special character of the example which we are considering.

5. LAGRANGE IDENTITY METHODS

The Lagrange identity method proceeds as follows: suppose we are able to construct a function V(x,t) which satisfies the adjoint equation (the forward heat equation) and homogeneous boundary conditions. Then by the Lagrange identity we have

$$\begin{aligned} 0 &= \int_o^t \int_D \left\{ V(x,\eta)\left[\frac{\partial u}{\partial \eta}(x,\eta) + \Delta u(x,\eta)\right] - u(x,\eta)\left[-\frac{\partial V}{\partial \eta}(x,\eta) + \Delta V(x,\eta)\right]\right\} dx d\eta \\ &= \int_D u(x,\eta)V(x,\eta)dx \Big|_o^t . \end{aligned} \tag{31}$$

Thus, for any such function $V(x,\eta)$ we obtain

$$\int_D u(x,t)V(x,t)dx = \int_D f(x)V(x,0)dx . \tag{32}$$

A particular choice of $V(x,\eta)$ which leads to interesting results is

$$V(x,\eta) = u(x,2t-\eta) . \tag{33}$$

Then clearly

$$V(x,t) = u(x,t)$$
$$V(x,0) = u(x,2t) \ , \tag{34}$$

and (32) becomes

$$\int_D u^2(x,t)dx = \int_D f(x)u(x,2t)dx \ . \tag{35}$$

In particular,

$$\int_D u^2(x,\tfrac{T}{2})dx \leq \left\{\int_D |u(x,0)|^p dx\right\}^{\frac{1}{p}}\left\{\int_D |u(x,T)|^q dx\right\}^{\frac{1}{q}}, \ \frac{1}{p}+\frac{1}{q} = 1$$
$$\equiv ||f(x)||_p ||u(x,T)||_q \tag{36}$$

and if we knew in addition that

$$||u(x,T)||_q \leq K_q \tag{37}$$

it would follow that

$$||u(x,\tfrac{T}{2})||_2 \leq K_q ||f(x)||_p \ . \tag{38}$$

One observes now the interesting fact that if the initial data is uniformly bounded and K_1 is prescribed then an application of (20) with T replaced by $\frac{T}{2}$ establishes Hölder continuous dependence in L_2 for $t \leq \frac{T}{2}$. On the other hand if $f \epsilon L_1$ and $|u(x,T)| \leq K_\infty$ then one establishes Hölder continuous dependence in L_2 for $0 < t < T$. (The same result could have been obtained in this special problem using eigenfunction expansion.)

This method has been used by Brun [13,14] in treating problems in elastodynamics, viscoelasticity and thermoelasticity. It can clearly be used to obtain explicit bounds for the solution (which is assumed to lie in the appropriate class).

6. METHODS OF PROTTER

The method of Protter gives a weak result for the simple problem (3) which we are considering. However, it is applicable to a variety of problems for which the other methods fail. Protter's idea is to set

$$u(x,t) = e^{\alpha t} w(x,t) \tag{39}$$

for an as yet undetermined constant α. Then w satisfies

$$\frac{\partial w}{\partial t} + \alpha w + \Delta w = 0 \qquad \text{in } D\times(0,T)\ . \tag{40}$$

We now form

$$0 = \int_o^t\int_D \left[\frac{\partial w}{\partial \eta} + \alpha w + \Delta w\right]^2 dx d\eta \geq 2\int_o^t\int_D \frac{\partial w}{\partial \eta}\,(\Delta w+\alpha w) dx d\eta\ . \tag{41}$$

An integration of the right hand side yields the inequality

$$\alpha||w(x,t)||_2^2 - ||\text{grad} w(x,t)||_2^2 \leq \alpha||f||_2^2 - ||\text{grad} f||_2^2\ . \tag{42}$$

Reinserting u(x,t) for w(x,t) one finds

$$||u(x,t)||_2^2 \leq \frac{1}{\alpha}||\text{grad} u(x,t)||_2^2 + \left[||f||_2^2 - \frac{||\text{grad} f||_2^2}{\alpha}\right] e^{\alpha t}\ . \tag{43}$$

We now assume that we know from observation that

$$||\text{grad} u||_2^2 \leq \tilde{K}^2 \tag{44}$$

for all $t\varepsilon(0,T)$. It then follows that

$$(u,u) \leq \frac{\tilde{K}^2}{\alpha} + (f,f)e^{\alpha t}\ . \tag{45}$$

If one now chooses

$$\alpha = \frac{1}{2T}\log\frac{1}{(f,f)} \tag{46}$$

then (45) becomes

$$(u,u) \leq \frac{2\tilde{K}^2 T}{\ln\left[\frac{1}{(f,f)}\right]} + (f,f)^{1-t/T}\ , \tag{47}$$

which yields logarithmic continuous dependence on the data in $0 \leq t < T$, a much weaker result than that obtained by logarithmic convexity or by the Lagrange identity method. The method has, however, been successfully applied to various types of improperly posed problems by Murray [105] and Murray and Protter [106], in some cases yielding results not obtainable by the other techniques discussed herein.

7. CONCLUDING REMARKS

We end this paper by listing some of the advantages and shortcomings of the various methods we have discussed. The eigenfunction methods tend to be less generally applicable than many of the other methods, and cannot be directly applied, for instance, to simple linear operator equations of the form

$$\frac{\partial u}{\partial t} = Mu ,$$

(where M is a linear operator on some appropriate Hilbert space) unless the eigenvalue problem

$$Mu = \lambda u$$

has a set of eigenfunctions which are complete in the appropriate sense. If M is a non-linear operator or a differential operator of no definite type the eigenfunction method will usually not be applicable.

Quasireversibility methods involve the definition of an approximate problem which is well posed and which yields a solution that in some sense is expected to be close to the solution of the original problem. There is no unique approximate problem and it may be difficult to show that

$$||U(x,0) - f(x)|| \leq \delta$$

for δ small, in more general non-linear operator equations. Some of the difficulties which arise in the application of this method are discussed in the paper of Miller [99] which appears in this volume.

Logarithmic convexity methods and similar methods which reduce the problem to one of solving a second order differential inequality, can be extended to handle a wide variety of non-linear operator equations and inequalities. Several such extensions are discussed in the accompanying papers of Knops [66] and Levine [87]. In general, this method seems to be restricted to operators in Hilbert space.

The Lagrange identity methods clearly allow us to move slightly away from this Hilbert space restriction. However, they do not appear to be easily applicable to non-linear operators and those which depend explicitly on the parameter t. In applications they have been restricted mainly to symmetric operators.

The method of Protter seems to be much better suited to equations involving higher derivatives with respect to t, e.g.

$$\frac{\partial^3 u}{\partial t^3} = Mu$$

and to certain classes of non-symmetric operators M. Murray [105] has recently used this method to extend the known linear elastodynamical uniqueness results.

Discussions of function theoretic methods and numerical methods are given in the papers of Colton [25] and Miller [99] which appear in this volume.

The accompanying bibliography is by no means complete. However, most of the research on improperly posed problems (of the classes mentioned herein) which has appeared in the literature prior to 1970 is referred to in the bibliographies of the various papers cited.

REFERENCES

[1] Abdul-Latif, A.I., Uniqueness of the solution of some hyperbolic boundary value problems, Ph.D. dissertation, University of Maryland (1966).

[2] Adelson, L., Singular perturbations of a class of improperly posed problems, Ph.D. dissertation, Cornell (1970).

[3] Agmon, S., Unique continuation and lower bounds for solutions of abstract differential equations, Proc. Inter. Congress Math. Stockholm (1962), pp. 301-305.

[4] Agmon, S., Unicité et convexité dans les problèmes différentiels, Sem. Math. Sup. (1965), University of Montreal Press (1966).

[5] Agmon, S., Lower bounds for solutions of Schrödinger equations, J. D'Analyse Math., 23 (1970), pp. 1-25.

[6] Agmon, S., and Nirenberg, L., Lower bounds and uniqueness theorems for solutions of differential equations in Hilbert space, Comm. Pure Appl. Math., 20 (1967), pp. 207-229.

[7] Aronszajn, N., A unique continuation theorem for solutions of elliptic partial differential equations or inequalities of second order, J. Math. Pure Appl., 36 (1957), pp. 235-249.

[8] Aziz, A.K., Gilbert, R.P., and Howard, H.C., A second order non-linear elliptic boundary value problem with generalised Goursat data, Ann. Mat. Pura Appl., 72 (1966), pp. 325-341.

[9] Bergman, S., Integral operators in the theory of linear partial differential equations, (Erg. d. math. u. Grenzgebiete, vol. 23) Springer, Berlin (1961).

[10] Bertolini, F., Sul problema di Cauchy per l'equazioné di Laplace in tre variabili independenti, Ann. Mat. Pura Appl., 40 (1956), pp. 121-128.

[11] Berzanskii, Ju. M., A uniqueness theorem in the inverse spectral problem for Schrödinger's equation, Trudy Mosk. Mat. Obsc., 7 (1968)

[12] Bourghin, D.G., and Duffin, R.J., The Dirichlet problem for the vibrating string equation, Bulletin Amer. Math. Soc., 45 (1939), pp. 851-858.

[13] Brun, L., Sur l'unicité en thermoélasticité dynamique et diverses expressions anologues à la formule de Clapeyron, C.R. Acad. Sci. Paris, 261 (1965), pp. 2584-2587.

[14] Brun, L., Méthodes energétiques dans les systèmes évolutifs linéaires, Premier partie: Séparation des énergies, Deuxième partie: Théoremès d'unicité, J. de Mechanique, 8 (1969), pp. 125-166, 167-192.

[15] Cahen, G., *Determination experimentale des parametres des systèmes a retard*, Revue Franc. Trait Inf. (1964), pp. 15-23.

[16] Calderon, A.P., *Uniqueness in the Cauchy problem for partial differential equations*, Amer. J. Math., 80 (1958), pp. 16-36.

[17] Cannon, J.R., *Determination of the unknown coefficient in a parabolic differential equation*, Duke Math. J., 30 (1963), pp. 313-323, see also *Determination of certain parameters in heat conduction problems*, J. Math. Anal. Appl., 8 (1964), pp. 188-201.

[18] Cannon, J.R., *A Cauchy problem for the heat equation*, Ann. Mat. Pura Appl., 66 (1964), pp. 155-165.

[19] Cannon, J.R., *Determination of the unknown coefficient k(u) in the equation* $\nabla \cdot k(u) \nabla u = 0$ *from overspecified boundary data*, J. Math. Anal. Appl., 18 (1967), pp. 112-114.

[20] Cannon, J.R., *Determination of an unknown heat source from over-specified boundary data*, SIAM J. Numer. Anal., 5 (1968), pp. 275-286.

[21] Cannon, J.R., and Dunninger, D.R., *Determination of an unknown forcing function in a hyperbolic equation from overspecified data*, Ann. Mat. Pura Appl., 85 (1970), pp. 49-62.

[22] Cannon, J.R., and Hill, C.D., *Continuous dependence of bounded solutions of a linear parabolic differential equation upon interior Cauchy data*, Duke Math. J., 35 (1968), pp. 217-230.

[23] Carleman, T., *Sur un problème de unicité pour des systèmes d'équations aux dérivées partielles à deux variables indépendantes*, Ark. Math. Astr. Fys., 26B (1939), pp. 1-9.

[24] Cohen, P.J., and Lees, M., *Asymptotic decay of solutions of differential inequalities*, Pac. J. Math., 11 (1961), pp. 1235-1249.

[25] Colton, D., Paper in this volume.

[26] Colton, D., *Cauchy's problem for almost linear elliptic equations in two independent variables*, J. Approx. Theory, 3 (1970), pp. 66-71.

[27] Conlan, J., and Trytten, G., *Pointwise bounds in the Cauchy problem for elliptic systems of partial differential equations*, Arch. Rat. Mech. Anal., 22 (1966), pp. 143-152.

[28] Cordes, H.O., *Über die Bestimmheit der Lösungen elliptischer Differentialgleichungen durch Anfangsvorgaben*, Nach. Akad. Wiss. Göttingen Math.-Phys., 11 (1956), pp. 239-258.

[29] Crooke, P.S., *On the Saffman model for the flow of a dusty gas and some related eigenvalue inequalities*, Ph.D., dissertation, Cornell (1970).

[30] Douglas, J., *The approximate solution of an unstable physical problem subject to constraints*, Functional Anal. and Optimization. Academic Press, New York (1966), pp. 65-66.

[31] Douglas, J., Approximate continuation of harmonic and parabolic functions, Numerical Sol. of Partial Diff. Eqtns., Academic Press (1966).

[32] Douglas, J., and Gallie, T.M., An approximate solution of an improper boundary value problem, Duke Math. J., 26 (1959), pp. 339-348.

[33] Douglas, J., and Jones, B.F., The determination of a coefficient in a parabolic differential equation, II Numerical approximation, J. Math. Mech., 11 (1962), pp. 919-926.

[34] Douglis, A., Uniqueness in the Cauchy problem for elliptic systems of equations, Comm. Pure Appl. Math., 6 (1953), pp. 291-298.

[35] Douglis, A., On uniqueness in Cauchy problems for elliptic systems of equations, Comm. Pure Appl. Math., 13 (1960), pp. 593-607.

[36] Dunninger, D.R., and Zachmonoglou, E.C., The condition for uniqueness of the Dirichlet problem for hyperbolic equations in cylindrical domains, J. Math. and Mech., 18 (1969), pp. 763-766.

[37] Dyer, R.H., Paper in this volume.

[38] Dyer, R.H., and Edmunds, D.E., Lower bounds for solutions of the Navier-Stokes equations, Proc. Lond. Math. Soc., 18 (1968), pp. 169-178.

[39] Edelstein, W.S., A uniqueness theorem in the linear theory of elasticity with microstructure, Acta. Mech., 8 (1969), pp. 183-184.

[40] Edmunds, D.E., Asymptotic behaviour of solutions of the Navier-Stokes equations, Arch. Rat. Mech. Anal., 22 (1966), pp. 15-21.

[41] Fadeev, L.D., Increasing solutions of Schrödinger's equation, Dokl. Akad. Nauk SSSR, 165 (1965), pp. 514-517.

[42] Fichera, G., Linear elliptic differential systems and eigenvalue problems, Lecture Notes in Math., No. 8, Springer-Verlag (1965).

[43] Foias, C., Gussi, G., and Poenaru, V., Sur le problème de Cauchy pour le type elliptique à deux variables, Acad. Rep. Pop. Roman. Bull. Mat. Fiz., 7 (1955), pp. 97-103.

[44] Fox, D.W., and Pucci, C., The Dirichlet problem for the wave equation, Ann. di Mat. Pura Appl., 46 (1958), pp. 155-182.

[45] Franklin, J., Well posed stochastic extensions of ill-posed linear problems, J. Math. Anal. Appl., 31 (1970), pp. 682-716.

[46] Gajewski, H., and Zacharias, K., Zur regularisierung liner Klasse nicht-korrekter Probleme bei Evolutionsgleichungen, J. Math. Anal. Appl., 38 (1972), pp. 784-789.

[47] Garabedian, P., Partial differential equations, Interscience Publishers, New York (1964).

[48] Garabedian, P., and Lieberstein, H.M., On the numerical calculation of detached bow shock waves in hypersonic flow, J. Aero. Sci., 25 (1958), pp. 109-118.

[49] Gilbert, R.P., Function theoretic methods in partial differential equations, Academic Press (1969).

[50] Hadamard, J., Lectures on the Cauchy problem in linear partial differential equations, Yale University Press (1923).

[51] Heinz, E., Über die Eindentigkeit beim Cauchyschen Anfangswertproblem einer elliptischen Differentialgleichung zweiter Ordnung, Nach. Akad. Wiss. Göttingen Math. Phys. IIa, 10 (1955), pp. 1-12.

[52] Hills, R., On the stability of linear dipolar fluids and On uniqueness and continuous dependence for a linear micropolar fluid, (both to appear)

[53] Holmgren, E., Über Systeme von linearen partielle Differentialgleichungen, Ofv. kongl. Vet.-Akad. Förk., 58 (1901), pp. 91-103.

[54] Hörmander, L., On the uniqueness of the Cauchy problem, Math. Scand., 6 (1958), pp. 213-225.

[55] Hyman, M., Electrostatic focussing of electron beams by coaxial cylinders, Symp. on Partial Diff. Eqtns. and Cont. Mech. Math. Research Center, University of Wisconsin (1960), p. 354.

[55a] Radley, D.E., The theory of the Pierce-type electron gun, J. Electr. Constr., 4 (1958), see also Electrodes for convergent Pierce-type electron guns, J. Electr. Constr., 15 (1963).

[56] Ivanov, V.K., The Cauchy problem for the Laplace equation in an infinite strip, Diff. Uravneniya, 1 (1965), pp. 131-136.

[57] Ivanov, V.K., On ill-posed problems, Math. 5b, 61 (1965), pp. 211-223, see also On a method in the theory of incorrect problems of mathematical physics, Symp. Partial Diff. Eqtns. (Novosibirsk) (1963), pp. 102-107.

[58] John, F., The Dirichlet problem for the wave equation, Amer. J. Math., 63 (1941), pp. 141-154.

[59] John, F., Numerical solution of the heat equation for preceding time, Ann. Mat. Pura Appl., 40 (1955), pp. 129-142.

[60] John, F., A note on "improper" problems in partial differential equations, Comm. Pure Appl. Math., 8 (1955), pp. 494-495.

[61] John, F., Numerical solution of problems which are not well-posed in the sense of Hadamard, Proc. Rome Symp. Prov. Int. Comp. Center (1959), pp. 103-116.

[62] John, F., Continuous dependence on data for solutions of partial differential equations with a prescribed bound, Comm. Pure Appl. Math., 13 (1960), pp. 551-585.

[63] Jones, B.F., The determination of a coefficient in a parabolic differential equation; I Existence and Uniqueness, J. Math. Mech., 11 (1962), pp. 907-918.

[64] Khosrovshahi, G.B., Growth properties for solutions of Schrödinger type systems, Ph.D. dissertation, Cornell (1972).

[65] Khosrovshahi, G.B., Paper in this volume.

[66] Knops, R.J., Paper in this volume.

[67] Knops, R.J., and Payne, L.E., Uniqueness in classical elastodynamics, Arch. Rat. Mech. Anal., 27 (1968), pp. 349-355.

[68] Knops, R.J., and Payne, L.E., Stability in linear elasticity, Int. J. Solids Structures, 4 (1968), pp. 1233-1242.

[69] Knops, R.J., and Payne, L.E., On the stability of solutions of the Navier-Stokes equations backward in time, Arch. Rat. Mech. Anal., 29 (1968), pp. 331-335.

[70] Knops, R.J., and Payne, L.E., Continuous data dependence for the equations of classical elastodynamics, Proc. Camb. Phil. Soc., 66 (1969), pp. 481-491.

[71] Knops, R.J., and Payne, L.E., On uniqueness and continuous data dependence in dynamical problems of linear thermoelasticity, Int. J. Structures Solid, 6 (1970), pp. 1173-1184.

[72] Knops, R.J., and Payne, L.E., Growth estimates for solutions of evolutionary equations in Hilbert space with applications in elastodynamics, Arch. Rat. Mech. Anal., 41 (1971), pp. 363-398.

[73] Knops, R.J., and Steel, T.R., Uniqueness in the linear theory of a mixture of two elastic solids, Int. J. Eng. Sci., 7 (1969), pp. 571-577.

[74] Krein, S.G., On some classes of correctly posed boundary value problems, Dokl. Akad. Nauk SSSR, 114 (1957), pp. 1162-1165.

[75] Kumano-go, H., On the uniqueness of the Cauchy problem and the unique continuation theorem for elliptic equations, Osaka Math. J., 14 (1962), pp. 181-212.

[76] Landis, E.M., Certain properties of equations of elliptic type, Dokl. Akad. Nauk SSSR, 107 (1956), pp. 640-643.

[77] Lattes, R., and Lions, J.L., The method of quasireversibility. Applications to partial differential equations, Amer. Elsevier Publ. Co. Inc., New York (1969).

[78] Lavrentiev, M.M., On the Cauchy problem for the Laplace equation, Izvest. Akad. Nauk SSSR, Ser. Math., 120 (1956), pp. 819-842.

[79] Lavrentiev, M.M., On the problem of Cauchy for linear elliptic equations of second order, Dokl. Akad. Nauk SSSR, 112 (1957), pp. 195-197.

[80] Lavrentiev, M.M., Some improperly posed problems in mathematical physics, Springer-Verlag, New York (1967).

[81] Lavrentiev, M.M., Romanov, V.G., and Vasiliev, V.G., Multidimensional inverse problems for differential equations, Lecture Notes in Mathematics, 167, Springer Verlag (1970).

[82] Lax, P.D., A stability theorem for solutions of abstract differential equations, and its application to the study of local behaviour of solutions of elliptic equations, Comm. Pure Appl. Math., 9 (1956), pp. 747-766.

[83] Levine, H.A., Logarithmic convexity and the Cauchy problem for some abstract second order differential inequalities, J. Diff. Eqtns., 8 (1970), pp. 34-55.

[84] Levine, H.A., On a theorem of Knops and Payne in dynamical linear thermoelasticity, Arch. Rat. Mech. Anal., 38 (1970), pp. 290-307.

[85] Levine, H.A., Logarithmic convexity, first order differential inequalities and some applications, Trans. Amer. Math. Soc., 152 (1970), pp. 299-319.

[86] Levine, H.A., Logarithmic convexity and the Cauchy problem for $P(t)u_{tt} + M(t)u_t + N(t)u = 0$ in Hilbert space, (to appear).

[87] Levine, H.A., Paper in this volume.

[88] Lewy, H., An example of a smooth linear partial differential equation without solution, Ann. of Math., 66 (1957).

[89] Lions, J.L., and Malgrange, B., Sur l'unicité retrograde dans les problèmes mixtes paraboliques, Math. Scand., 8 (1960), pp. 277-286.

[90] Lopatinskii, Y.B., Uniqueness of the solution of Cauchy's problem for a class of elliptic equations, Dopov. Akad. Nauk Ukrain (1958), pp. 689-693.

[91] Malgrange, B., Unicité du problème de Cauchy d'après A.P. Calderon, Seminaire Bourbaki (1959), p. 178.

[92] Marčuk, G.I., On the formulation of certain inverse problems, Dokl. Akad. Nauk SSSR, 156 (1964), pp. 503-506.

[93] Marčuk, G.I., Certain problems in numerical and applied mathematics, Novosibirsk (1966).

[94] Miller, K., Three circle theorems in partial differential equations and applications to improperly posed problems, Arch. Rat. Mech. Anal., 16 (1964), pp. 126-154.

[95] Miller, K., An eigenfunction expression method for problems with overspecified data, Ann. Scuola Norm Sup di Pisa, 19 (1965), pp. 397-405.

[96] Miller, K., Stabilized numerical methods for location of poles by analytic continuation, Studies in Numerical Analysis 2: Numerical Solutions of Nonlinear Problems, Symposium SIAM, Philadelphia (1968), pp. 9-20, published SIAM (1970).

[97] Miller, K., Stabilized numerical analytic prolongment with poles, SIAM J. Appl. Math., 18 (1970), pp. 346-363.

[98] Miller, K., Least square methods for ill-posed problems with a prescribed bound, SIAM J. Math. Anal., 1 (1970), pp. 52-74.

[99] Miller, K., Paper in this volume.

[100] Miranker, W.L., A well-posed problem for the backward heat equation, Proc. Amer. Math. Soc., 12 (1961), pp. 243-247.

[101] Mizel, V., and Seidman, T.I., Observation and prediction for the heat equation, J. Math. Anal. Appl., 28 (1969), pp. 303-312, see also same journal, 38 (1972), pp. 149-166.

[102] Mizohata, S., Unicité dans le problème de Cauchy pour quelques equations differentielles elliptiques, Mem. Coll. Sci. Univ. Kyoto, 31 (1958), pp. 121-128.

[103] Mizohata, S., Unicité du prolongment des solutions des equations elliptiques du quatrième ordre, Proc. Jap. Acad., 34 (1958), pp. 687-692.

[104] Muller, C., On the behaviour of the solutions of the differential equation $\Delta u = F(x,u)$ in the neighbourhood of a point, Comm. Pure Appl. Math., 7 (1954), pp. 505-514.

[105] Murray, A., Uniqueness and continuous dependence for the equations of elastodynamics without strain energy function, Arch. Rat. Mech. Anal., 47 (1972), pp. 195-204.

[106] Murray, A., and Protter, M., Asymptotic behaviour of solutions of second order systems of partial differential equations, (to appear).

[106a] Naghdi, P.M., and Trapp, J.A., A uniqueness theorem in the theory of Cosserat surface, J. Elast., 2 (1972), pp. 9-20.

[107] Nirenberg, L., Uniqueness in Cauchy problems for differential equations with constant leading coefficients, Comm. Pure Appl. Math., 10 (1957), pp. 89-105.

[108] Novikov, P.S., On the inverse problem of potential theory, Dokl. Akad. Nauk SSSR, 18 (1938), pp. 165-168.

[109] Ogawa, H., Lower bounds for solutions of hyperbolic inequalities, Proc. Amer. Math. Soc., 16 (1965), pp. 853-857.

[110] Ogawa, H., Lower bounds for solutions of differential inequalities in Hilbert space, Proc. Amer. Math. Soc., 16 (1965), pp. 853-857.

[111] Ogawa, H., Lower bounds for the solutions of parabolic differential inequalities, Can. J. Math., 19 (1967), pp. 667-672.

[112] Payne, L.E., Bounds in the Cauchy problem for the Laplace equation, Arch. Rat. Mech. Anal., 5 (1960), pp. 35-45.

[113] Payne, L.E., On some non-well-posed problems for partial differential equations, Numer. Sol. of Nonlinear Diff. Eqtns., M.R.C. Conference University of Wisconsin, Wiley Press (1966), pp. 239-263.

[114] Payne, L.E., On a priori bounds in the Cauchy problem for elliptic equations, SIAM J. Math. Anal., 1 (1970), pp. 82-89.

[115] Payne, L.E., and Sather, D., On some non-well-posed problems for the Chaplygin equation, Math. Anal. Appl., 19 (1967), pp. 67-77.

[116] Payne, L.E., and Sather, D., On some non-well-posed Cauchy problems for quasilinear equations of mixed type, Trans. Amer. Math. Soc., 128 (1967), pp. 135-141.

[117] Payne, L.E., and Sather, D., On singular perturbations in non-well-posed problems, Ann. Mat. Pura Appl., 75 (1967), pp. 219-230.

[118] Payne, L.E., and Sather, D., On an initial-boundary value problem for a class of degenerate elliptic operators, Ann. Mat. Pura Appl., 78 (1968), pp. 323-338.

[119] Pederson, R.N., On the unique continuation formula for certain second and fourth order elliptic equations, Comm. Pure Appl. Math., 11 (1958), pp. 67-80.

[120] Pederson, R.N., Uniqueness and Cauchy's problem for elliptic equations with double characteristics, Ark. Mat., 6 (1967), pp. 535-549.

[121] Petrowsky, I.G., Lectures in partial differential equations, Interscience Publishers, New York (1954).

[122] Picone, M., Maggiorazione degli integrali delle equazioni totalmente paraboliche alle derivate parziali del secondo ordine, Ann. Mat. Pura Appl. (1929).

[123] Plis, A., Unique continuation theorems for solutions of partial differential equations, Proc. Int. Cong. Math. Stockholm (1962), pp. 397-402, see also Nonuniqueness in Cauchy's problem for differential equations of elliptic type, J. Math. Mech., 9 (1960).

[124] Plis, A., A smooth linear elliptic equation without any solution in a sphere, Comm. Pure Appl. Math., 14 (1961), pp. 599-617.

[125] Protter, M.H., Unique continuation for elliptic equations, Trans. Amer. Math. Soc., 95 (1960), pp. 81-91.

[126] Protter, M.H., Properties of solutions of parabolic equations and inequalities, Can. J. Math., 13 (1961), pp. 331-345.

[127] Protter, M.H., Asymptotic behaviour and uniqueness theorems for hyperbolic operators, Joint Symp. Part. Diff. Eqtns. (Novosibirsk 1963) Acad. Sci. USSR Moscow (1963), pp. 348-353.

[128] Pucci, C., Studio col metodo delle differenze di un problema di Cauchy relativo ad equazioni a derivate parziali del secondo ordine di tipo parabolico, Ann. Scuola Norm Pisa, 7 (1953), pp. 205-215.

[129] Pucci, C., Sui problemi di Cauchy non "ben posti", Rend. Acad. Naz. Lincei, 18 (1955), pp. 473-477.

[130] Pucci, C., Discussione del problema di Cauchy per le equazioni di tipo ellitico, Ann. Mat. Pura Appl., 46 (1958), pp. 131-153.

[131] Pucci, C., Some topics in parabolic and elliptic equations, Lecture Series, No. 36, Institute for Fluid Dynamics and Appl. Math., University of Maryland (1958).

[132] Pucci, C., Discussione del problema di Cauchy per le equazioni di tipo ellitico, Ann. Math. Pura Appl., 46 (1958), pp. 391-412.

[133] Rappaport, I.M., On a two dimensional inverse problem in potential theory, Dokl. Akad. Nauk SSSR, 19 (1938).

[134] Sather, D., and Sather, J., The Cauchy problem for an elliptic parabolic operator, Ann. Mat. Pura Appl., 30 (1968), pp. 197-214.

[135] Schaefer, P.W., On the Cauchy problem for an elliptic system, Arch. Rat. Mech. Anal., 20 (1965), pp. 391-412.

[136] Schaefer, P.W., On the Cauchy problem for the nonlinear biharmonic equation, J. Math. Anal. Appl., 36 (1971), pp. 660-673.

[137] Shirota, T., A remark on the unique continuation theorem for certain fourth order elliptic equations, Proc. Jap. Acad., 36 (1960), pp. 571-573.

[138] Sretenskii, L.N., On an inverse problem of potential theory, Izv. Akad. Nauk SSSR, Ser. Math., 2 (1938).

[139] Tihonov, A.N., On stability of inverse problems, Dokl. Akad. Nauk SSSR, 39 (1944), pp. 195-198.

[140] Tihonov, A.N., On the solution of ill-posed problems and the method of regularisation, Dokl. Akad. Nauk SSSR, 151 (1963), pp. 501-504.

[141] Tihonov, A.N., On the regularisation of ill-posed problems, Dokl. Akad. Nauk SSSR, 153 (1963), pp. 49-52.

[142] Trytten, G.N., Pointwise bounds for solutions of the Cauchy problem for elliptic equations, Arch. Rat. Mech. Anal., 13 (1963), pp. 222-244.

[143] Vekua, I.N., New methods for solving elliptic equations, Gos. Izdat. Tech. Teor. Lit., Moscow (1948).

[144] Zimmerman, J.M., Band limited functions and improper boundary value problems for a class of nonlinear partial differential equations, J. of Math. Mech., 11 (1962), pp. 183-196.

LOGARITHMIC CONVEXITY AND OTHER TECHNIQUES APPLIED TO PROBLEMS IN CONTINUUM MECHANICS

R.J. KNOPS

1. INTRODUCTION

In the opening lecture of the symposium, Professor Payne has described several techniques for the analysis of solutions to non-well-posed problems. To illustrate these techniques, he considered the heat conduction equation backwards in time. Here, we are primarily concerned with the further application of the techniques of logarithmic convexity and related convexity arguments, Lagrange identities, and the Protter-Murray weighted energy method to equations of continuum mechanics. We wish to establish conditions under which the solution is unique and depends continuously (in a certain sense) on its data. In addition, for certain non-well-posed problems, we will obtain estimates for the growth (in time) of the solution measured by a suitable norm. Some other equations will be considered in detail by later speakers and other aspects will also be taken up by them. In particular, Levine will discuss generalisations of arguments used in this talk to weak solutions of abstract equations of parabolic and hyperbolic type. In a broad sense, therefore, this and the later contributions supplement and amplify the opening lecture by Payne.

It is impossible in a short space to present computations for all the problems treated by the techniques just mentioned, and so details will be given for only a selection. These, however, will be drawn mainly either from recently developed theories or from non-linear theories of continuum mechanics.

2. LOGARITHMIC CONVEXITY

Convexity arguments have been in use for some considerable time, yet their application to problems in mathematical physics, including continuum mechanics, is comparatively recent and may be said to have started with the work of Pucci, John and Lavrentiev and continued with that of Agmon, Nirenberg and Payne and his co-workers. (See the bibliographies contained in the first lecture and in the paper by Payne [39].)

We shall briefly sketch the ideas underlying the technique of logarithmic convexity, referring the interested reader to the comprehensive account by Agmon [1] for a detailed treatment. Afterwards, we consider the implications of this technique for uniqueness, Hölder continuity and growth estimates.

It is well known that a function $f(t) \varepsilon C^2(0,T)$ is convex if it satisfies the differential inequality

$$f''(t) \geqslant 0 , \tag{2.1}$$

where a prime denotes differentiation w.r.t. the variable t. Geometrically, the curve representing a convex function must (i) be below the straight line segment joining any two points on the curve; and (ii) be above the tangent drawn at any point to the curve. (see Figures 1 and 2)

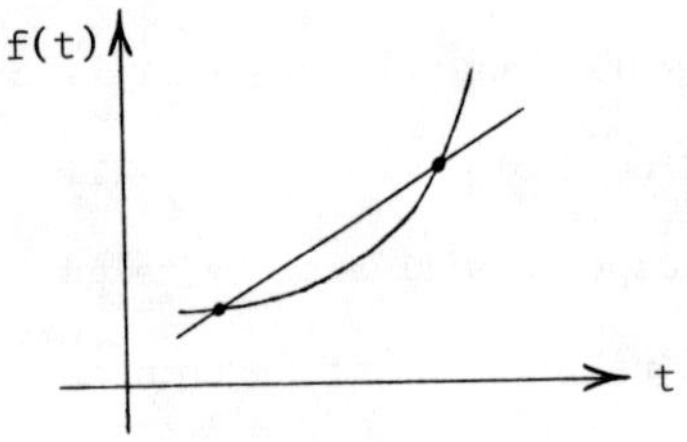

Figure 1

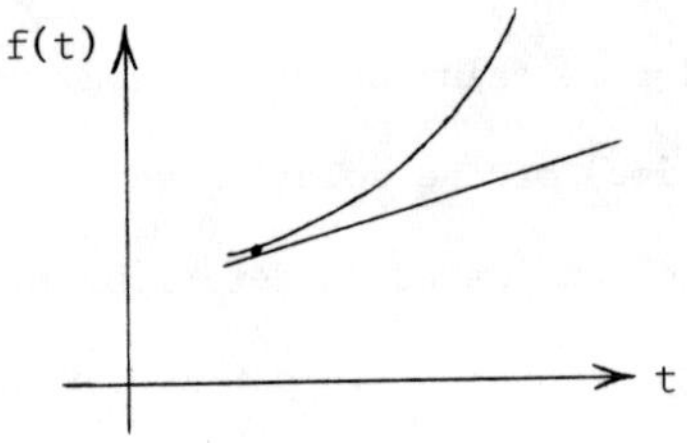

Figure 2

The analytic counterparts of these simple geometric properties form the basis of several subsequent calculations.

Because the problems that we will treat from continuum mechanics occur as partial differential equations, it cannot be expected that they will naturally yield a function satisfying an ordinary differential inequality of the kind (2.1). Instead, we should expect a O.D inequality in some (Banach) space X of the solutions, with f not necessarily equal to the norm defined on X, but rather associated with this norm by some functional relationship. In fact, slightly more general situations can be handled, in which the norm can be replaced by a certain positive-definite function F, and the right-side of (2.1) replaced by various combinations of F and its first derivative. Moreover, it turns out that only two special kinds of functional dependence of f upon F have so far produced any practical results. They are the logarithmic and polynomial cases and therefore attention is restricted to these, with polynomial dependence being discussed in the next section. As a general rule, logarithmic dependence is useful in linear problems and in those with non-linearities due to convective terms (e.g. Navier-Stokes equations).

Thus, for the technique of logarithmic convexity, we must construct a real-valued twice continuously differentiable function F(t) defined on the solutions $u(\cdot,t)$ (1), and possessing the following two properties

$$F(t) \geq 0 , \quad 0 < t < T \qquad \text{(i)}$$

$$F(t) = 0 \Longleftrightarrow u(\cdot,t) = 0, \quad 0 < t < T . \qquad \text{(ii)}$$

Then, by virtue of the governing equations and data, we must prove that F(t) satisfies a differential inequality of the kind

$$FF'' - (F')^2 \geq - a_1 FF' - a_2 F^2 , \tag{2.2}$$

where a_1, a_2 are constants. Different forms for the right-side of (2.2) are also possible (see for instance Khosrovshahi [14]); in particular, the case where a_1, a_2 are functions of time may be treated (see for example Agmon [1]). It is evident that the precise form of F(t) is dictated by the particular problem under study.

---

(1) For the rigorous definition of F(t) on the set of solutions, see the lectures by Levine and Miller in this volume.

To see that inequality (2.2) is a generalisation of (2.1), let us assume that $F(t) > 0$, $0 < t_1 < t < t_2 < T$. Then, (2.2) becomes

$$(\ell n F(t))'' + a_1(\ell n F)' + a_2 \geqslant 0, \qquad t_1 < t < t_2 , \tag{2.3}$$

which, for $a_1 = a_2 = 0$, is the desired result after setting $f(t) = \ell n F(t)$. Otherwise, the change of variable

$$\sigma = e^{-a_1 t} \tag{2.4}$$

transforms (2.3) into

$$\frac{d^2}{d\sigma^2} \ell n \left[F(\sigma)\sigma^{-a_2/a_1^2} \right] \geqslant 0, \qquad \sigma_1 < \sigma < \sigma_2 , \tag{2.5}$$

$$\sigma_\alpha = e^{-a_1 t_\alpha}, \qquad \alpha = 1,2 ,$$

which clearly has the same form as (2.1). Thus, $\ell n F(t) - a_2 t/a_1$ is a convex function of $e^{-a_1 t}$. In view of the reduction of (2.2) to (2.5), the presence of a non-trivial right-side to (2.2) only slightly affects the continuity of the solution on its data, and the uniqueness of the solution not at all. However, the presence of such terms is highly significant for questions of growth.

To prove uniqueness and continuous data dependence, we integrate (2.3) by means of Jensen's inequality (equivalent to the geometric property depicted in Figure 1) to obtain

$$F(t) \leqslant e^{-\frac{a_2}{a_1} t} \left[F(t_1) e^{\frac{a_2}{a_1} t_1} \right]^{\delta} \left[F(t_2) e^{\frac{a_2}{a_1} t_2} \right]^{1-\delta} , \quad 0 \leqslant t_1 < t < t_2 \leqslant T, \tag{2.6}$$

where

$$\delta = \frac{e^{-a_1 t} - e^{-a_1 t_2}}{e^{-a_1 t_1} - e^{-a_1 t_2}} . \tag{2.7}$$

Uniqueness is established by demonstrating that $F(0) = 0$ implies $F(t) = 0, 0 \leq t \leq T$. Assume the contrary, so that $F(t) > 0$ on $t \varepsilon (t_1, t_2) \subset [0,T]$. Then either $t_1 = 0$ or, by continuity, $F(t_1) = 0$. If $F(t_1) = 0$ then using (2.6), it may be shown (Agmon [1] and Levine [26]) that $F(t) = 0$, $t \varepsilon [0,t_2)$, and hence by continuity that $F(t) = 0$, $t \varepsilon [0,T]$. When $t_1 = 0$, we have $F(0) = 0$, and the same conclusion follows by repetition of the argument. Uniqueness is therefore established.

As indicated in the introductory lecture, continuous dependence of the solution $u(\cdot,t)$ upon its (initial) data can be established only in a restricted class of solutions, namely those which at time T satisfy

$$F(T) \leq Me^{-a_2 T/a_1}, \tag{2.8}$$

for some positive constant M. Initial data is measured by $F(0)$, required to be small, but not zero, and so by virtue of the uniqueness just proved, we may assume $F(t) > 0$, $t \varepsilon [0,T]$. Thus, (2.6) is valid on the interval $[0,T]$, and, on using (2.8), may be expressed in the form

$$F(t) \leq kM^{1-\delta}[F(0)]^{\delta}, \quad 0 \leq \delta(t) \leq 1, \tag{2.9}$$

for some constant $k(T)$. Inequality (2.9) clearly shows that the solution depends <u>Hölder continuously</u> on its initial data <u>in the measure F</u>, on compact sub-intervals of $[0,T]$. (See Pucci [40], John [12] and [13].) This type of continuity must not be confused with continuous dependence in the usual sense nor with Liapounov stability. (These concepts correspond respectively with continuity on the closed finite interval $[0,T]$ and on the semi-infinite interval $[0,\infty)$.) Thus, in the case of well-posed problems or for problems with Liapounov stable solutions, the use of (2.9) and Hölder continuity become superfluous. On the other hand, in those non-well-posed problems or problems with unstable solutions which admit a function $F(t)$ with a convex logarithm, (2.8) and (2.9) demonstrate how the imposition of a priori bounds lead to the recovery of continuous dependence, in the modified sense of Hölder. Such questions have been generally discussed by F. John [13], who also proposed the concept of logarithmic continuity which we shall introduce later in connexion with the Protter-Murray method.

While the topic of continuous dependence is clearly very important, it is also of interest to learn how the solutions to non-well-posed and related problems evolve with time. Information, in the form of estimates, may be obtained by integrating the fundamental inequality (2.2) in the manner analogous to the second geometrical property depicted in Figure 2. For simplicity, we deal first with the case when $a_1 = a_2 = 0$. Then, from (2.2) (under the assumption that $F(0) \not\equiv 0$) we obtain

$$F(t) \geq F(0) \exp\left\{t \frac{F'(0)}{F(0)}\right\}, \qquad t \geq 0, \tag{2.10}$$

so that when $F'(0) \geq 0$, we see that F(t) is bounded below by an increasing exponential function of time. When $F'(0) = 0$, we have

$$F(t) \geq F(0). \tag{2.11}$$

Let us return to (2.6) and write it as

$$F(t) \leq F(0) \exp\left\{\frac{t}{T} \ln \frac{F(T)}{F(0)}\right\}, \qquad 0 \leq t \leq T. \tag{2.12}$$

If we assume that $\lim_{T\to\infty} \frac{1}{T} \ln F(T) = 0$ (or alternatively that $F(T) < \beta e^{\gamma T^{1-\varepsilon}} T^N$, where β, γ, ε are positive constants and N is any positive integer) then from (2.12) we see that

$$F(t) \leq F(0), \qquad t \geq 0. \tag{2.13}$$

Thus, by virtue of (2.11) and (2.13), solutions with $F'(0) = 0$ and with the asymptotic behaviour just described, must satisfy

$$F(t) = F(0).$$

Arguments of similar kind may be used to furnish other conclusions. We collect all the results in the following theorem (for proof, see for instance Knops and Payne [19] and [21]).

Theorem: Let the function $F(t)$ have a convex logarithm. Then

(i) either (a) $F(t)$ is bounded below by a time-increasing exponential function for sufficiently large time,

or (b) $F(t) \lesssim F(0)$;

(ii) when $F'(0) > 0$, $F(t)$ is bounded below by a time-increasing exponential function for $t \geq 0$;

(iii) when $F'(0) = 0$

either (c) $F(t)$ is bounded below by a time-increasing exponential function for sufficiently large time,

or (d) $F(t) = F(0)$;

(iv) when $F'(0) < 0$, then either (i)(a) or (i)(b) are possible.

In general, the ambiguity in (i), (iii), (iv) cannot be removed without the introduction of further assumptions.

Now let us consider the case when a_1, a_2 are non-zero. Integration of (2.2), using the "tangent property" of Figure 1, leads to

$$F(t) \geq F(0) \exp\left[\left(\frac{F'(0) + \frac{a_2}{a_1} F(0)}{a_1 F(0)}\right)\left(1-e^{-a_1 t}\right) - \frac{a_2}{a_1} t\right],$$

from which it easily follows that $F(t)$ is bounded below by a time-increasing exponential provided

either (i) $a_1 > 0$, $a_2 < 0$,

or (ii) $a_1 < 0$, $a_1 F'(0) + a_2 F(0) < 0$.

The same result also holds for $a_1 = 0$, $a_2 > 0$.

Generalisations and refinements of the above arguments are evidently possible, and details of some of these may be found in Agmon [1], Ogawa [36], [37], [38], Levine [27], Levine, Knops and Payne [32]; see also Khosrovshahi [14].

It is worthwhile remarking that the above theorem and corresponding results for non-zero a_1, a_2 demonstrate the non-existence of the class of smooth solutions for which a possible F(t) may be defined. In particular, we see that when the convexity inequality is satisfies, there are no solutions with a polynomial growth behaviour for large values of time.

3. EXAMPLES

Rather than use the theory of linearised anisotropic elasticity[(1)] to illustrate the remarks of the preceding section, we shall use instead a more recent example from the director theory of rods (see Green, Knops and Laws [9]).

Consider a curve embedded in euclidean three-space, which at each point P has associated with it two assigned vectors, called directors, and consider three configurations of the rod: the initial configuration in which the directors are denoted by $\bar{\underset{\sim}{A}}_\alpha(\theta)$ $(\alpha = 1,2)$ and the position vector of P is denoted by $\bar{\underset{\sim}{R}}(\theta)$, where θ is a convected coordinate; a deformed equilibrium configuration in which the directors are denoted by $\underset{\sim}{A}_\alpha(\theta)$ and the position vector of P by $\underset{\sim}{R}(\theta)$; and a final configuration in which the directors are denoted by $\underset{\sim}{a}_\alpha$ and the position vector of P by $\underset{\sim}{a}_\alpha(\theta,t)$. We further define

$$\bar{\underset{\sim}{A}}_3 = \frac{\partial \bar{\underset{\sim}{R}}}{\partial \theta}, \qquad \underset{\sim}{A}_3 = \frac{\partial \underset{\sim}{R}}{\partial \theta}, \qquad \underset{\sim}{a}_3 = \frac{\partial \underset{\sim}{r}}{\partial \theta}.$$

Let us assume that the final deformation is small in the sense that

$$\underset{\sim}{a}_\alpha = \underset{\sim}{A}_\alpha + \varepsilon \underset{\sim}{b}_\alpha,$$

$$\underset{\sim}{r} = \underset{\sim}{R} + \varepsilon \underset{\sim}{u},$$

and orders of ε higher than the first can be neglected. Let us define $\underset{\sim}{b}_3$ and b_{ij} through

$$\underset{\sim}{b}_3 = \frac{\partial \underset{\sim}{u}}{\partial \theta}, \qquad \underset{\sim}{b}_i = b_{ij}\underset{\sim}{A}_j .$$

---

(1) Logarithmic convexity has been applied to this theory by Knops and Payne [15], [16], [18], [19], [22].

It may be shown (see Green, Knops and Laws [9]) that for a simply extended director rod of length ℓ, in which $\bar{A}_i$, A_i are orthonormal sets, the equations of motion for flexure in the plane normal to A_1 are

$$(1-\xi\lambda^2)v_{,xx} + b_{,x} = m\ddot{v} , \tag{3.1}$$

$$0 \leq x \leq 1 ,$$

$$\xi b_{,xx} - b - v_{,x} = n\ddot{b} , \tag{3.2}$$

where a dot now indicates differentiation w.r.t. t,

$$\theta = \ell x , \quad \ell v = u_2 , \quad b = b_{23} ,$$

and ξ, m, n are certain material constants, which may be assumed positive. The parameter λ^2 is proportional to the end load, and is positive or negative according as the rod is in compression or tension. Finally, for the purposes of this example, we assume the rod has clamped ends so that

$$v(t,x) = b(t,x) = 0 , \quad \text{at } x = 0,1 . \tag{3.3}$$

The following conservation law is an immediate consequence of (3.1)-(3.3)

$$E(t) = \tfrac{1}{2}\int_0^1 \left[n\dot{v}^2+n\dot{b}^2+(1-\xi\lambda^2)(v_{,x})^2+2bv_{,x}+\xi(b_{,x})^2+b^2\right]dx = E(0) . \tag{3.4}$$

For the function F(t) we take

$$F(t) = \int_0^1 (mv^2+nb^2)dx + \beta(t+t_0)^2 , \tag{3.5}$$

where β, t_0 are positive constants to be determined later. We suppose a classical solution exists to (3.1)-(3.3) and then (3.5) satisfies the differentiability requirements; clearly it is also positive-definite.

To show that F(t) given by (3.5) satisfies an inequality of the kind (2.2), we differentiate (3.5) and use the equations of motion (3.1), (3.2) to substitute for the inertia terms arising in the expression for the second time differential. Thus,

$$F'(t) = 2\int_0^1 (mv\dot{v}+nb\dot{b})dx + 2\beta(t+t_0) ,$$

$$F''(t) = 2\int_0^1 (m\dot{v}^2+n\dot{b}^2)dx - 2\int_0^1\left[(1-\xi\lambda^2)v^2_{,x}+2bv_{,x}+\xi b^2_{,x}+b^2\right]dx + 2\beta$$

$$= 4\int_0^1 (m\dot{v}^2+n\dot{b}^2)dx - 4E(0) + 2\beta ,$$

where in the last line use has been made of (3.4). It may now easily be shown by means of Schwarz's inequality that

$$FF'' - (F')^2 \geqslant - 2(2E(0)+\beta)F(t) \ , \tag{3.6}$$

which is of the desired form. Uniqueness and continuous data dependence then follow from the remarks of Section 2. (Observe that for uniqueness, the initial data is given by $v(x,o) = \dot{v}(x,o) = b(x,o) = \dot{b}(x,o) = 0$ and therefore on setting $\beta = 0$, we have $F(0) = 0$.)

Growth of $F(t)$ may be easily established in the case $2E(0) + \beta \leqslant 0$ (i.e. when the initial total energy $E(0)$ is negative) as then the right-side of (3.6) is positive and can therefore be dropped. Hence, we may obtain the integrated form (2.10) in which, however, because of the present form (3.5) of $F(t)$, we may always make $F'(0) > 0$ by choosing t_o suitably large. We conclude at once, therefore, that

$$G(t) \equiv \int_o^1 (mv^2+nb^2)dx \tag{3.7}$$

has a time increasing exponential lower bound for sufficiently large time.

When $E(0) = 0$, we choose $\beta = 0$, and (3.6) now corresponds to (2.2) with $a_1 = a_2 = 0$. Discussion of growth estimates under these conditions has already been given in Section 2.

When $E(0) > 0$, we cannot expect growth in all situations since it is known that when the strain-energy is positive-definite the null solution is Liapounov stable. However, provided this latter condition is violated, growth of $G(t)$ may be established if, for instance, $G'(0) \geqslant 2\left[2G(0)E(0)\right]^{\frac{1}{2}}$. Further details may be found in the general treatment of Knops and Payne [21].

It must be remarked that growth, for this particular example, has been established only in terms of the measure (3.7), and therefore it is uncertain whether growth is due to an increase in either the v- or b-components of $G(t)$, or indeed whether both components oscillate unboundedly and exactly "out-of-phase" for large values of time. This remains still largely an open question, although E.W. Wilkes and the author have obtained some partial results.

The general method of analysis outlined above for the director rod theory holds generally for weak solutions to equations of the form

$$M\ddot{u} = Nu \,, \tag{3.8}$$

where M, N are linear, time-independent, symmetric operators with values in a Hilbert space $\mathcal{H}$, and M is also positive-definite. An appropriate choice for F(t) is

$$F(t) = ||(\cdot)||^2_{\mathcal{H}} + \beta(t+t_o)^2 \,,$$

where $||(\cdot)||_{\mathcal{H}}$ is the norm defined on $\mathcal{H}$. A general discussion is given by Knops and Payne [21], but special cases of (3.8) appear for example in linearised theories of elasticity and multipolar elasticity (see Edelstein [7]), and also in the fields of plasma physics (Laval, Mercier and Pellat [24]) and stellar dynamics (Antonov [3]). It is interesting to note that Laval, Mercier and Pellat quite independently in 1955 developed the technique of logarithmic convexity to discuss certain instabilities arising in their problem.

One possible generalisation of (3.8) is via the introduction of dissipation

$$M\ddot{u} + D\dot{u} = Nu \,. \tag{3.9}$$

Levine [26] has shown that (3.9) specialises to the equations of linearised thermoelasticity (when u = (w,θ), where w is the three-dimensional vector displacement and θ is the incremental scalar temperature). Dafermos [6] has studied the asymptotic behaviour of the solutions to equation (3.9) and has shown that another specialisation of them corresponds to the linearised theory of two interacting homogeneous isotropic elastic materials (see Steel [41]).

It is possible to apply the technique of logarithmic convexity to equations of the form (3.9), and prove that, under suitable conditions, appropriately defined functions F satisfy inequalities of the type

$$FF'' - (F')^2 \geq - a_1F^2 - a_2FF' , \tag{3.10}$$

where a_1, a_2 are constants whose sign depends upon the precise problem under consideration. For instance, in thermoelasticity (see Levine [26], Knops and Payne [21]), we may take

$$F(t) = \int_o^t\int_{B(\eta)} \rho w_i w_i dx d\eta + (T-t)\int_{B(o)} \rho w_i w_i dx + \gamma , \tag{3.11}$$

where w_i is the displacement, γ is some computable positive constant, ρ is the density, and $B(\eta)$ denotes integration over the volume B of the body at time t. For the mixture of two linear elastic materials (see Knops and Steel [22], [23]), we may take

$$F(t) = \int_o^t\int_{B(\eta)} (\rho_1 w_i w_i d + \rho_2 v_i v_i) dx d\eta \tag{3.12}$$

where w_i, v_i are the displacements of the constituent materials, and ρ_1, ρ_2 their densities.

As an example of a non-linear system of equations, we mention the incompressible Navier-Stokes equation backward in time in which logarithmic convexity arguments have been used to establish uniqueness and continuous dependence on the data (Knops and Payne [17]). Similar arguments may be applied to the Navier-Stokes equation forward in time. Hölder continuous dependence of the solution upon its initial data may be established by assuming that two solutions $u_i^{(1)}$, $u_i^{(2)}$ exist for separate initial data

$f_i^{(1)}$, $f_i^{(2)}$, and that these solutions are restricted to satisfy

$$\sup_{x,t} u_i^{(1)} u_i^{(1)} \leq M^2 , \tag{3.13}$$

$$\sup_{x,t} \left[u_i^{(2)} u_i^{(2)} + \left(u_{i,j}^{(2)}-u_{j,i}^{(2)}\right)\left(u_{i,j}^{(2)}-u_{j,i}^{(2)}\right) + u_i^{(2)} u_i^{(2)}\right] \leq N^2 , \tag{3.14}$$

for positive constants M, N. Then on setting $v_i = u_i^{(1)} - u_i^{(2)}$ it may be shown, with the help of equations governing v_i, easily derived from (3.13), that the function

$$F(t) = \int_{B(t)} v_i v_i dx$$

satisfies an inequality of the type (3.10) with $a_1 < 0$, $a_2 < 0$. Uniqueness and continuous dependence then follow as described in Section 2. By taking

$$F(t) = \int_t^o \int_{B(\eta)} v_i v_i dx d\eta + (t_o+t) \int_{B(o)} v_i v_i + \gamma ,$$

where γ is a positive constant, uniqueness and continuous dependence may again be established by logarithmic convexity, but under assumptions weaker than those given by (3.14).

Logarithmic convexity has also been applied by Hills [10], [11] to the linear dipolar fluid and linear micropolar fluid to establish uniqueness and continuous dependence. The governing equations of motion are non-linear due to the presence of convective terms.

4. FURTHER CONVEXITY ARGUMENTS

The method that will be described in this section, developed mainly by Levine, Payne and Knops, is closely allied to the techniques of logarithmic convexity discussed previously but uses instead the polynomial functional relation between f and F (see Section 2) and moreover examines consequences of <u>concavity inequalities</u> of the kind

$$(F^{\gamma})'' \leq 0 .$$

Generally, this approach is applicable to non-linear theories of both parabolic and hyperbolic type (see for instance Levine [27], Levine and Payne [29], [30], Levine, Knops and Payne [32]), but we shall content ourselves in this article with treating the single example of isothermal (non-linear) elasticity with a strain-energy function, W.

Thus, let $u_i(x_i,t)$ be the displacement at time t of a point whose position vector in the reference configuration B is x_i. Then the Piola-Kirchhoff stress σ_{ij} satisfies the equations of motion

$$\sigma_{ij,j} = \rho_o \ddot{u}_i , \tag{4.1}$$

where ρ_o, assumed positive, is the mass density in the reference configuration, and the constitutive relations

$$\sigma_{ij} = \frac{\partial W}{\partial u_{i,j}} , \qquad W = W(u_{i,j}) \tag{4.2}$$

where W is the strain energy function per unit volume of the elastic body in its reference configuration. Homogeneous data are assigned on the surface

$$\begin{aligned} u_i &= 0 \text{ on } \overline{\partial B_1} \\ n_j \sigma_{ij} &= 0 \text{ on } \partial B_2 \end{aligned} \tag{4.3}$$

where $\overline{\partial B_1} \cup \partial B_2 = \partial B$, and $\partial B_2 \neq \phi$; n_j is the unit normal on ∂B_2.

It easily follows from (4.1)-(4.3) that

$$E(t) \equiv \tfrac{1}{2} \int_B \rho_o \dot{u}_i \dot{u}_i dx + \int_B W dx = E(0) . \tag{4.4}$$

We now postulate that there exists a constant $\alpha > 2$ such that(1)

$$\int_B \left\{ \alpha W - u_{i,j} \frac{\partial W}{\partial u_{i,j}} \right\} dx \gtrless 0 , \tag{4.5}$$

and by means of (4.1)-(4.5) prove that

$$F(t) = \int_B \rho_o u_i u_i dx + \beta (t+t_o)^2 , \qquad 0 < t < T \tag{4.6}$$

(1) When $\alpha = 2$, the subsequent calculations breakdown and must be replaced by those of logarithmic convexity. Note that linearised elasticity satisfies (4.5) identically with $\alpha = 2$.

is <u>polynomially concave</u> on the interval $(0,T)$, where, as before, β and t_o are positive constants to be suitably chosen. Thus,

$$F'(t) = 2\int_B \rho_o u_i \dot{u}_i dx + 2\beta(t+t_o)$$

and

$$F''(t) = (2+\alpha)\int_B \rho_o \dot{u}_i \dot{u}_i dx - 2\alpha E(0) + 2\int_B \left(\alpha W - u_{i,j}\frac{\partial W}{\partial u_{i,j}}\right) dx + 2\beta$$

$$\geqslant (2+\alpha)\int_B \rho_o \dot{u}_i \dot{u}_i dx - 2\alpha E(0) + 2\beta .$$

We may now use Schwarz's inequality to show that

$$FF'' - \left(\frac{2+\alpha}{4}\right)(F')^2 \geqslant -\alpha(\beta+2E(0))F . \tag{4.7}$$

Since $\alpha > 2$, expression (4.7) has the equivalent form

$$(F^{-\gamma})'' \leqslant 2\gamma(1+2\gamma)F^{-(\gamma+1)}(\beta+2E(0)), \qquad \gamma = \frac{\alpha-2}{4} > 0 , \tag{4.8}$$

<u>provided $F(t) > 0$</u>.

Treatment of (4.8) follows closely that of the corresponding logarithmic convexity inequality. For instance, when initial data and the form of W are such that $E(0) \leqslant 0$, the choice of β making $2E(0) + \beta = 0$, enables the right-side of (4.8) to be omitted. Jensen's inequality then shows that

$$F^{\gamma}(t) \leqslant \frac{F^{\gamma}(0)F^{\gamma}(T)}{(1-t/T)F^{\gamma}(T)+(t/T)F^{\gamma}(0)} \tag{4.9}$$

from which it may be deduced that the null solution to the initial boundary value problem (4.1)-(4.3) is both unique and depends Hölder continuously on its initial data, provided (4.5) holds(1).

To establish estimates for the growth of solutions, we integrate (4.8) according to the "tangent" property (c.p. Figure 2) to obtain

$$F^{\gamma}(t) \geqslant F^{\gamma}(0)/\left[1-\gamma t\,\frac{F'(0)}{F(0)}\right] , \tag{4.10}$$

(1) Alternatively, (4.5) may be regarded as restricting the class of admissible displacement yielding uniqueness, continuous dependence, etc.

where again we have assumed $\beta + 2E(0) < 0$. Irrespective of the initial data, t_o can be chosen such that $F'(0) > 0$, and hence (4.10) proves that $F(t)$ becomes unbounded in finite time. This result contrasts with linearised elasticity where solutions, in general, become unbounded only after infinite time, but is in keeping with other non-linear hyperbolic theories (see Levine [27]).

Further conclusions are possible when initial data and the form of W are such that $E(0) > 0$, but they are somewhat too complicated to describe here.

5. LAGRANGE IDENTITIES

We illustrate this technique by applying it to the initial mixed boundary value of linearised elasticity (see Brun [4], [5]). Other applications in continuum mechanics are to linearised thermoelasticity (Brun [4], [5], Green [8]), viscoelasticity (Brun [4], [5]) and to the linearised theory of Cosserat surfaces (Naghdi and Trapp [35]).

The equations of linear elasticity are

$$(c_{ijkl}u_{k,l})_{,j} = \rho \ddot{u}_i \text{ in } B\times(0,T) , \tag{5.1}$$

in which u_i is the small displacement, $\rho = \rho(x) > 0$ is the mass density, and $c_{ijkl}(x_i)$ are the elasticities assumed to satisfy the symmetry conditions

$$c_{ijkl} = c_{klij} . \tag{5.2}$$

The last condition makes (5.1) formally self-adjoint. (Self-adjointness is not essential to the success of the method; see the opening lecture by Payne.) As boundary conditions, we adopt the homogeneous data

$$\begin{aligned} u_i &= 0 \text{ on } \overline{\partial B_1}\times(0,T) \\ n_j c_{ijkl}u_{k,l} &= 0 \text{ on } \partial B_2\times(0,T) , \end{aligned} \tag{5.3}$$

where $\partial B = \overline{\partial B_1} \cup \partial B_2$. Initially, the values of $u_i(x,o)$ and $\dot{u}_i(x,o)$ are assigned and we suppose that a classical solution exists to the initial boundary value problem (5.1)-(5.3).

Now, for any two sufficiently differentiable (distinct) functions $v_i(x,t)$, $w_i(x,t)$, the following identity holds.

$$\int_{B(t)} \rho v_i(x,t)\dot{w}_i(x,t)dx = \int_o^t \int_{B(\eta)} [\rho \dot{v}_i(x,\eta)\dot{w}_i(x,\eta) + \rho v_i(x,\eta)\ddot{w}_i(x,\eta)]dxd\eta$$

$$+ \int_{B(0)} \rho v_i(x,o)\dot{w}_i(x,o)dx , \qquad (5.4)$$

where again B(t) denotes integration over the volume of the body at time t. Let us set

$$w_i(x,t) = u_i(x,t) ,$$

$$v_i(x,\eta) = - u_i(x,2t-\eta) .$$

Then, by means of (5.1)-(5.3) and integration by parts, (5.4) becomes

$$\int_{B(t)} \rho u_i \dot{u}_i dx = \tfrac{1}{2} \int_B [\rho u_i(x,o)\dot{u}_i(x,2t) + \rho \dot{u}_i(x,o)u_i(x,2t)]dx . \qquad (5.5)$$

Uniqueness of the solution to the initial boundary value problem may be immediately deduced from (5.5) on setting $u_i(x,o) = \dot{u}_i(x,o) = 0$. We observe that only the symmetry condition (5.2) is required on the elasticities, the present uniqueness proof not requiring definiteness conditions usually found in classical proofs. This result and method of proof are due to Brun [4], [5], but it is worth recording that the same result may be obtained either using logarithmic convexity (Knops and Payne [15]) or the method of Protter and Murray [33] (see also Murray [34]). In the next section, uniqueness in the same problem will be established by the latter method for skew-symmetric elasticities.

Hölder continuous dependence on the initial data in an L_2-norm sense may be deduced from (5.5) in the manner described in the introductory lecture by Payne. The fundamental identity (5.5) may also be treated slightly differently when the initial data admits $\dot{u}_i(x,o) = 0$. An integration w.r.t. time and use of the arithmetic-geometric mean inequality then leads to

$$G(t) \leq \frac{3}{4} G(0) + \frac{1}{4} G(2t) \leq \frac{3}{4} \sum_{i=1}^{N-1} \frac{1}{4^i} G(0) + \frac{1}{4^N} G(2^N t) , \tag{5.6}$$

for any integral N, where $G(t) = \int_{B(t)} \rho u_i u_i dx$. Thus, on letting $N \to \infty$, we see that solutions with the asymptotic behaviour

$$G(t) = O(t^2) \text{ as } t \to \infty , \tag{5.7}$$

must remain less than or equal to their initial value in the sense that

$$G(t) \leq G(0) , \quad t \geq 0 . \tag{5.8}$$

An alternative choice of the functions v_i and w_i is

$$\begin{aligned} w_i(x,t) &= u_i(x,t) \\ v_i(x,\eta) &= \frac{\partial u_i}{\partial \eta} (2t-\eta) . \end{aligned} \tag{5.9}$$

Substitution into (5.4) and use of (5.1) then gives

$$\begin{aligned} \int_{B(t)} \rho \dot{u}_i \dot{u}_i dx &= \int_o^t \int_{B(\eta)} \left[c_{ijkl} \dot{u}_{i,j}(\eta) u_{k,l}(2t-\eta) - c_{ijkl} \dot{u}_{i,j}(2t-\eta) u_{k,l}(\eta) \right] dx d\eta \\ &\quad + \int_B \rho \dot{u}_i(2t) \dot{u}_i(0) dx \\ &= \int_{B(t)} c_{ijkl} u_{i,j}(t) u_{k,l}(t) d\eta - \int_B c_{ijkl} u_{i,j}(0) u_{k,l}(2t) dx \\ &\quad + \int_B \rho \dot{u}_i(2t) \dot{u}_i(0) dx . \end{aligned}$$

Elimination of the strain energy by means of the energy conservation equation finally produces

$$\int_{B(t)} \rho \dot{u}_i \dot{u}_i dx = \tfrac{1}{2} V(0) + \tfrac{1}{2} T(0) - \tfrac{1}{2} \int_B c_{ijkl} u_{i,j}(0) u_{k,l}(2t) dx + \tfrac{1}{2} \int_B \rho \dot{u}_i(2t) \dot{u}_i(0) dx$$

where V, T are the strain and kinetic energies, respectively. For simplicity we suppose $u_i(x,o) = 0$, so that

$$T(t) = \tfrac{1}{2}T(0) + \tfrac{1}{2}\int_B \rho\dot{u}_i(2t)\dot{u}_i(0)dx$$

$$\leq \frac{3}{4}\,T(0) + \frac{1}{4}\,T(2t)\ ,$$

and in a manner analogous to that used in treating (5.6) we see that if $T(t) = O(t^2)$ as $t \to \infty$, then $T(t) \leq T(0)$, $t \geq 0$. By means of the identity

$$G(t) = G(0) + tG'(0) + 8\int_o^t (t-\eta)T(\eta)d\eta - 2E(0)t^2\ ,$$

this in turn implies $G(t) \leq t^2T(0)$ and hence $G(t) = O(t^2)$ as $t \to \infty$.

We remark that results of this section do not require the coefficients c_{ijkl} to be sign-definite. All the results, of course, carry over for abstract operators of the form $M\ddot{u} = Nu$ mentioned in Section 3.

6. THE PROTTER-MURRAY TECHNIQUE

For purposes of illustration, we again consider the example of linearised elasticity introduced in the previous section, with, however, the elasticities now being skew-symmetric

$$c_{ijkl} = -\,c_{klij}\ . \tag{6.1}$$

Murray [34] has considered elasticities more general than those satisfying either (5.2) or (5.1), while Levine has extended the treatment to certain abstract equations.

To apply the Protter-Murray technique, we start by defining the function $w(x,t)$ according to

$$w_i(x,t) = e^{-\lambda t}u_i(x,t)\ , \tag{6.2}$$

where λ (>0) is a parameter. Substitution of (6.2) into the equations of motion (5.1) yields the following equations for w

$$(c_{ijkl}w_{k,l})_{,j} - \rho(\ddot{w}_i + 2\lambda\dot{w}_i + \lambda^2 w_i) = 0\ . \tag{6.3}$$

Because $2ab \leq (a+b)^2$, appropriate partition of terms in (6.3) leads at once to

$$\int_B \rho\left\{\ddot{w}_i+\lambda^2 w_i\right\}\left\{2\rho\lambda\dot{w}_i-(c_{ijkl}w_{k,l})_{,j}\right\}dx \leq 0 ,$$

or

$$\lambda \frac{d}{dt}\left[\int_B \rho^2(\dot{w}_i\dot{w}_i dx+\lambda^2 w_i w_i)dx\right] - \frac{d}{dt}\int_B \rho(c_{ijkl}w_{k,l})_{,j}\dot{w}_i dx \leq 0 ,$$

where (6.1) has been employed.

Integration then gives

$$\lambda\int_{B(t)} \rho^2\dot{w}_i\dot{w}_i dx + \lambda^3\int_{B(t)} \rho^2 w_i w_i dx - \int_{B(t)} \rho(c_{ijkl}w_{k,l})_{,j}w_i dx$$
$$\leq \lambda\int_{B(o)} \rho^2\dot{w}_i\dot{w}_i dx + \lambda^3\int_{B(o)} \rho^2 w_i w_i dx + \int_{B(o)} \rho c_{ijkl}w_{k,l}\dot{w}_{i,j} dx . \tag{6.4}$$

We now express (6.4) in terms of u_i by means of (6.2) and obtain

$$\lambda\int_{B(t)} \rho^2(\dot{u}_i-\ u_i)(u_i-\ u_i)dx + \lambda^3\int_{B(t)} \rho^2 u_i u_i dx$$
$$-\int_{B(t)} \rho(c_{ijkl}u_{k,l})_{,j}(\dot{u}_i-\lambda u_i)dx \leq e^{2\lambda t}Q , \tag{6.5}$$

where the initial data term Q is given by

$$Q \equiv \lambda\int_{B(o)} \rho^2(\dot{u}_i-\lambda u_i)(\dot{u}_i-\lambda u_i)dx+\lambda^3\int_{B(o)} \rho^2 u_i u_i dx+\int_{B(o)} \rho c_{ijkl}u_{k,l}(\dot{u}_{i,j}-\lambda u_{i,j})dx$$
$$\leq 2\lambda^3\alpha_1\int_{B(o)} u_i u_i dx+(\lambda+1)\alpha_1\int_{B(o)} \dot{u}_i\dot{u}_i dx+\alpha_2\int_{B(o)} (c_{ijkl}u_{k,l})_{,j}(c_{ipqr}u_{q,r})_{,p}dx , \tag{6.6}$$

for computable constants α_1, α_2. We now follow a development due to Levine, and employ the inequality, for constant $\gamma > 0$,

$$\int \rho(c_{ijkl}u_{k,l})_{,j}(\dot{u}_i-\lambda u_i)dx \leq \frac{\gamma}{2}\int (c_{ijkl}u_{k,l})_{,j}(c_{ipqr}u_{q,r})_{,p}dx$$
$$+\frac{1}{2\gamma}\int \rho^2(\dot{u}_i-\lambda u_i)(\dot{u}_i-\lambda u_i)dx ,$$

in the further reduction of (6.5) to

$$\lambda^3 \int_{B(t)} \rho^2 u_i u_i dx - \frac{1}{4\lambda} \int_{B(t)} (c_{ijkl} u_{k,l})_{,j} (c_{ipqr} u_{q,r})_{,p} dx \leq e^{2\lambda t} Q \ . \quad (6.7)$$

We now deduce from (6.7) the required properties of uniqueness and continuous dependence in the class $\mathfrak{M}$ of classical solutions given by

$$\mathfrak{M} = \left\{ u : \sup_{0 \leq t \leq T} \int_{B(t)} (c_{ijkl} u_{k,l})_{,j} (c_{ipqr} u_{q,r})_{,p} dx \leq M^2 \right\} , \quad (6.8)$$

for some positive constant M. For uniqueness, we take homogeneous initial data, so that $Q \equiv 0$, and then for each value of $t \varepsilon [0,T]$ we let $\lambda \to \infty$. Immediately from (6.7) we obtain $u_i(x,t) \equiv 0$, and uniqueness follows.

To obtain continuous dependence, we put

$$I^2 = \int_{B(o)} u_i u_i dx + \int_{B(o)} \dot{u}_i \dot{u}_i dx$$

and select λ according to

$$2\lambda T = - \ell n I \ , \quad (6.9)$$

where $0 \leq t \leq T$. Because I is defined in terms of the initial data, we may assume, for the purposes of the present objectives, that $0 \leq I < 1$. For solutions in the class $\mathfrak{M}$ we then see that (6.7) yields

$$\int_{B(t)} \rho^2 u_i u_i dx \leq \frac{2T}{-\ell n I} \left[- \frac{2MT^3}{(\ell n I)^3} + \frac{4MT^2}{I(\ell n I)^2} - \alpha_1 \frac{2TI}{(+\ell n I)} + \alpha_1 \frac{4T^2 I}{(\ell n I)^2} - \alpha_1 \frac{I \ell n I}{2T} \right] ,$$

giving the required continuous dependence.

The original proof of Murray [34], relying on hypotheses slightly different to those adopted here, produces an inequality different to (6.8), but which yields the same general kind of dependence. Of course, a body-force may always be included in the calculations.

Other applications of the technique to problems in continuum mechanics are currently being investigated by several people, including Hills.

REFERENCES

[1] Agmon, S., Unicité et convexité dans les problèmes différentiels, Sem. Math. Sup (1965), University of Montreal Press (1966).

[2] Agmon, S., and Nirenberg, L., Lower Bounds and uniqueness theorems of differential equations in a Hilbert space, Comm. Pure Appl. Math., 20 (1967), pp. 207-229.

[3] Antonov, V.A., Remarks on the problem of stability in Stellar dynamics, Sov. Astr., 4 (1961), pp. 859-867 ≡ Ast. Zh., 37 (1960), pp. 918-926.

[4] Brun, L., Sur l'unicité en thermoélasticité dynamique et diverses expressions anologues à la formule de Clapeyron, C.R. Acad. Sci. Paris, 261 (1965), pp. 2584-2587.

[5] Brun, L., Méthodes energétiques dans les systèmes èvolutifs linéaires, Premier partie: Séparation des énergies; Deuxième partie: Théoremès d'unicité, J. de Mechanique, 8 (1969), pp. 125-166, 167-192.

[6] Dafermos, C.M., Wave equations with weak damping, SIAM J. Appl. Math., 18 (1970), pp. 759-767.

[7] Edelstein, W.S., A uniqueness theorem in the linear theory of elasticity with microstructure, Acta. Mech., 8 (1969), pp. 183-184.

[8] Green, A.E., (to appear in Mathematika).

[9] Green, A.E., Knops, R.J., and Laws, N., Large deformations, superposed small deformations and stability of elastic rods, Int. J. Sols. Structs., 4 (1968), pp. 555-577.

[10] Hills, R.N., On the stability of linear dipolar fluids, (to appear).

[11] Hills, R.N., On uniqueness and continuous dependence for a linear micropolar fluid, (to appear).

[12] John, F., A note on "improper" problems in partial differential equations, Comm. Pure Appl. Math., 8 (1955), pp. 494-495.

[13] John, F., Continuous dependence on data for solutions of partial differential equations with a prescribed bound, Comm. Pure Appl. Math., 13 (1960), pp. 551-585.

[14] Khosrovshahi, G.B., (to appear).

[15] Knops, R.J., and Payne, L.E., Uniqueness in classical elastodynamics, Arch. Rat. Mech. Anal., 27 (1968), pp. 349-355.

[16] Knops, R.J., and Payne, L.E., Stability in linear elasticity, Int. J. Solids Structures, 4 (1968), pp. 1233-1242.

[17] Knops, R.J., and Payne, L.E., On the stability of solutions of the Navier-Stokes equations backward in time, Arch. Rat. Mech. Anal., 29 (1968), pp. 331-335.

[18] Knops, R.J., and Payne, L.E., Continuous data dependence for the equations of classical elastodynamics, Proc. Camb. Phil. Soc., 66 (1969), pp. 481-491.

[19] Knops, R.J., and Payne, L.E., Hölder stability and logarithmic convexity, IUTAM Symposium on Instability of Continuous Systems, Herrenalb (1969).

[20] Knops, R.J., and Payne, L.E., On uniqueness and continuous data dependence in dynamical problems of linear thermoelasticity, Int. J. Structures Solid, 6 (1970), pp. 1173-1184.

[21] Knops, R.J., and Payne, L.E., Growth estimates for solutions of evolutionary equations in Hilbert space with applications in elastodynamics, Arch. Rat. Mech. Anal., 41 (1971), pp. 363-398.

[22] Knops, R.J., and Steel, T.R., Uniqueness in the linear theory of a mixture of two elastic solids, Int. J. Eng. Sci., 7 (1969), pp. 571-577.

[23] Knops, R.J., and Steel, T.R., On the stability of a mixture of two elastic solids, J. Comp. Mats., 3 (1969), pp. 652-663.

[24] Laval, G., Mercier, C., and Pellat, R., Necessity of the energy principles for magnetostatic stability, Nuclear Fusion, 5 (1965), pp. 156-158.

[25] Levine, H.A., Logarithmic convexity and the Cauchy problem for abstract second order differential inequalities, J. Diff. Eqns., 8 (1969), pp. 34-55.

[26] Levine, H.A., On a theorem of Knops and Payne in dynamical linear thermoelasticity, Arch. Rat. Mech. Anal., 38 (1970), pp. 290-307.

[27] Levine, H.A., Instability and nonexistence of global solutions to non-linear wave equations of the form $Pu_{tt} = - Au + \mathcal{F}(u)$. Lecture Notes, University of Dundee (1972).

[28] Levine, H.A., Logarithmic convexity and the Cauchy problem for $P(t)u_{tt} + M(t)u_t + N(t)u = 0$ in Hilbert space, (to appear).

[29] Levine, H.A., and Payne, L.E., Instability and nonexistence of global solutions to non-linear evolutionary equations of the form $Pu_{tt} = A(\mathcal{F}(Au))$ and some examples, (to appear).

[30] Levine, H.A., and Payne, L.E., Instability, growth and nonexistence of global solutions to non-linear evolutionary equations of the form $Pu_t = A(\mathcal{F}(Au))$ and some examples, (to appear).

[31] Levine, H.A., Paper in this volume.

[32] Levine, H.A., Knops, R.J., and Payne, L.E., (to appear).

[33] Murray, A., and Protter, M., Asymptotic behaviour of solutions of second order systems of partial differential equations, (to appear).

[34] Murray, A., Uniqueness and continuous dependence for the equations of elastodynamics without strain energy function, (to appear).

[35] Naghdi, P.M., and Trapp, J.A., A uniqueness theorem in the theory of Cosserat surface, J. Elast., 2 (1972), pp. 9-20.

[36] Ogawa, H., Lower bounds for solutions of hyperbolic inequalities, Proc. Amer. Math. Soc., 16 (1965), pp. 853-857.

[37] Ogawa, H., Lower bounds for solutions of differential inequalities in Hilbert space, Proc. Amer. Math. Soc., 16 (1965), pp. 853-857.

[38] Ogawa, H., Lower bounds for the solutions of parabolic differential inequalities, Can. J. Math., 19 (1967), pp. 667-672.

[39] Payne, L.E., On some non-well-posed problems for partial differential equations, Numerical Solutions of Non-Linear Differential Equations, Wiley (1964), pp. 239-263.

[40] Pucci, C., Sui problemi di Cauchy non "ben posti", Rend. Acad. Naz. Lincei, 18 (1955), pp. 473-477.

[41] Steel, T.R., Applications of a theory of interacting continua, Quart. J. Mech. Appl. Math., 20 (1967), pp. 57-72.

CAUCHY'S PROBLEM AND THE ANALYTIC CONTINUATION OF SOLUTIONS TO ELLIPTIC EQUATIONS *

DAVID COLTON

Department of Mathematics
Indiana University
Bloomington, Indiana

and

Department of Mathematics
University of Glasgow
Glasgow, Scotland

I. INTRODUCTION

Although long ignored by mathematicians as merely a pathological example of a non-well posed problem, Cauchy's problem for elliptic equations nevertheless arises in a natural manner in several situations of physical interest. In particular such a problem can occur in the study of various inverse problems of mathematical physics. We illustrate this by the following three problems:

PROBLEM 1: To construct a steady irrotational flow of an incompressible fluid with a prescribed constant pressure free surface S. If we let $u(\underset{\sim}{x})$, $\underset{\sim}{x} = (x_1, \ldots, x_n)$, be the velocity potential, then this problem can be mathematically formulated as the following (one parameter) Cauchy problem for Laplace's equation: Find a function $u(\underset{\sim}{x})$ satisfying

$$\Delta_n u = 0 \quad ; \quad \underset{\sim}{x} \in \mathbb{R}^n \text{ .... singular points of } u$$

and satisfying the Cauchy data

$$\frac{\partial u}{\partial \nu} = 0 \quad ; \quad \underset{\sim}{x} \in S$$

$$|\Delta u|^2 = 1 \quad ; \quad \underset{\sim}{x} \in S$$

where S is a prescribed analytic surface and ν is the positive normal to S (c.f. [7]).

PROBLEM 2: Let D be a clamped membrane vibrating with frequency ω and bounded by a simple closed analytic curve S. Let c be the velocity of sound. Suppose we ask the following question: What must the initial velocity and displacement be in order to have a prescribed (analytic) slope of displacement along (a portion of)

---

* This research was supported in part by NSF Grant GP-27232 and in part by the Science Research Council while the author was a visiting research fellow at the University of Glasgow.

S? This question leads to the problem of finding a function $u(x,y)$ satisfying

$$\Delta_2 u + \frac{\omega^2}{c^2} u = 0 \quad ; \quad (x,y) \in D$$

$$u = 0 \quad ; \quad (x,y) \in S$$

$$\frac{\partial u}{\partial \nu} = g(x,y) \quad ; \quad (x,y) \in S$$

where ν is the inner normal to S and $g(x,y)$ is a prescribed analytic function.

PROBLEM 3: Let $u(\underset{\sim}{x})$ be a solution of the reduced wave equation

$$\Delta_3 u + \lambda^2 u = 0$$

in the exterior of a bounded domain D and let $u(\underset{\sim}{x})$ satisfy the Sommerfeld radiation condition

$$\lim_{r \to \infty} r\left(\frac{\partial u}{\partial r} - i\lambda u\right) = 0$$

where $r = |\underset{\sim}{x}|$. As r tends to infinity $u(\underset{\sim}{x})$ has the asymptotic representation (c.f. [12])

$$u(\underset{\sim}{x}) = \frac{e^{i\lambda r}}{r} f(\theta,\phi) + O\left(\frac{1}{r^2}\right)$$

where (r,θ,ϕ) are spherical coordinates. The function $f(\theta,\phi)$ is called the radiation pattern and uniquely determines $u(\underset{\sim}{x})$ in the exterior of some sphere S containing D in its interior. A basic problem in scattering theory is to analytically continue $u(\underset{\sim}{x})$ past the sphere S (c.f. [14]). By expanding $u(\underset{\sim}{x})$ in a series of spherical harmonics one can determine Cauchy data for $u(\underset{\sim}{x})$ on S, and hence the problem of analytic continuation is reduced to finding the maximum domain of regularity of the solution to a Cauchy problem for the reduced wave equation (For the solution of this problem in the case when $u(\underset{\sim}{x})$ is axially symmetric see [1] and [4]).

Each of the above problems can of course be solved <u>locally</u> by appealing to the Cauchy-Kowalewski theorem. However this is clearly unsuitable for our purposes (see also [13]) since what is required is a <u>global</u> solution of the elliptic Cauchy problem i.e. we are faced with the problem of analytically continuing the local solution obtained (for example) via the Cauchy-Kowalewski theorem. The following example shows that in order to preform such a continuation it is necessary to examine the behaviour of the initial data in the space of (possibly several) complex variables:

<u>Example</u>: The function

$$u(x,y) = \mathrm{Re}\left\{\frac{1}{1+z^2}\right\}$$

$$= \frac{1 + x^2 - y^2}{(1+x^2-y^2)^2 + 4x^2y^2}$$

is a harmonic function for $(x,y) \neq (o,\pm 1)$ which satisfies the Cauchy data

$$u(x,o) = \frac{1}{1+x^2}$$

$$\frac{\partial u(x,o)}{\partial y} = o \quad .$$

The singularity of $u(x,y)$ at the points $(x,y) = (o,\pm 1)$ arises from the fact that the analytic continuation of the initial data $u(x,o)$ into the complex domain has a singularity at $z = \pm i$.

In this lecture we will give procedures for constructing global solutions to the Cauchy problem

$$\Delta_n u + f(\underset{\sim}{x})u = o \quad ; \quad n = 2,3 \qquad (1.1)$$

$$u(\underset{\sim}{x}) = \phi(\underset{\sim}{x}) \quad ; \quad \underset{\sim}{x} \in S$$

$$\frac{\partial u(\underset{\sim}{x})}{\partial \nu} = \psi(\underset{\sim}{x}) \quad ; \quad \underset{\sim}{x} \in S \qquad (1.2)$$

where $f(\underset{\sim}{x})$ is an entire function of its independent (complex) variables and ν is the positive normal to the analytic surface S. For related approaches to this problem the reader is referred to the work of Henrici ([9]), Garabedian ([5], [6]), and Hill ([10]).

II. THE ELLIPTIC CAUCHY PROBLEM IN TWO SPACE VARIABLES

We now consider the Cauchy problem (1.1), (1.2) for the case $n=2$. We make the assumption that the analytic arc S is contained in a domain D which is conformally symmetric with respect to S, i.e. there exists a conformal mapping which transforms S into an interval of the real axis (containing the origin) and D into a domain G which is symmetric with respect to the real axis. Preforming such a conformal transformation and then making the nonsingular transformation of the space C^2 of two complex variables into itself defined by

$$z = x + iy$$
$$z^* = x - iy \qquad (2.1)$$

transforms the Cauchy problem (1.1), (1.2) into an initial value problem of the form ([2])

$$U_{zz^*} + F(z,z^*)U = 0 \qquad (2.2)$$

$$U(z,z^*) = \Phi(z) \quad ; \quad z = z^*$$

$$\frac{\partial U(z,z^*)}{\partial z} - \frac{\partial U(z,z^*)}{\partial z^*} = -i\,\Psi(z) \; ; \; z=z^* \qquad (2.3)$$

where $F(z,z^*)$ is analytic in $G \times G$ and $\Phi(z)$ and $\Psi(z)$ are analytic in G. Now let

$$\frac{\partial^2 U(z,z^*)}{\partial z \partial z^*} = \rho(z,z^*) \quad . \qquad (2.4)$$

Then $U(z,z^*)$ has the integral representation

$$U(z,z^*) = \underset{\sim}{B}(\rho) = \int_0^z \int_0^{z^*} \rho(\xi,\xi^*)d\xi^* d\xi - \int_0^z \int_0^{\xi} \rho(\xi,\xi^*)d\xi^* d\xi$$

$$- \int_0^{z^*} \int_0^{\xi^*} \rho(\xi,\xi^*)d\xi d\xi^* + \tfrac{1}{2}\int_0^z \left[\Phi'(\xi) - i\Psi(\xi)\right] d\xi \qquad (2.5)$$

$$+ \tfrac{1}{2}\int_0^{z^*} \left[\Phi'(\xi^*) + i\Psi(\xi^*)\right] d\xi^* + \Phi(0) \; .$$

The initial value problem (2.2), (2.3) can now be reduced to finding a fixed point of the operator equation

$$\rho = - F \underset{\sim}{B}(\rho) = \underset{\sim}{T}(\rho) \qquad (2.6)$$

in the Banach space HB of function of two complex variables which are holomorphic and bounded in G x G with respect to the norm

$$||\rho||_\lambda = \sup_{G \times G} \{ e^{-\lambda(|z| + |z^*|)} |\rho(z,z^*)| \} , \qquad (2.7)$$

where $\lambda > o$ is a fixed real number. From estimates of the form

$$\left| \int_o^z \rho(\xi,z^*)d\xi \right| \leq \int_o^{|z|} ||\rho||_\lambda e^{\lambda|\xi|+\lambda|z^*|} |d\xi|$$

$$\leq \frac{1}{\lambda} e^{\lambda|z|+\lambda|z^*|} ||\rho||_\lambda$$

i.e.

$$\left|\left| \int_o^z \rho(\xi,z^*)d\xi \right|\right|_\lambda \leq \frac{||\rho||_\lambda}{\lambda} , \qquad (2.8)$$

it is easily shown ([2]) that the operator $\underset{\sim}{T}$ is a contraction mapping for λ sufficiently large and takes a closed ball of HB into itself. Hence by the Banach contraction mapping theorem there exists a unique analytic solution ρ of the equation (2.6), and equation (2.5) now defines the unique solution of the initial value problem (2.2), (2.3). We summarize the above analysis in the following theorem:

<u>THEOREM:</u> Let D be a domain which is conformally symmetric with respect to the analytic arc S and let $\phi(x)$ and $\psi(x)$ be the restriction to the arc S of functions which are analytic in D. Then the solution of the Cauchy problem (1.1), (1.2) (for n=2) is

an analytic function of z and z^* for $(z,z^*) \varepsilon D \times D^*$ where $D^* = \{z^*: \overline{z^*} \varepsilon D\}$ and can be constructed by iteration.

III. THE ELLIPTIC CAUCHY PROBLEM IN THREE SPACE VARIABLES

We now consider the Cauchy problem (1.1), (1.2) for the case $n = 3$. We make the assumption that S is a convex analytic surface whose normal is never parallel to the x_1- axis. The change of variables

$$\begin{aligned} x &= x_1 \\ z &= x_2 + ix_3 \\ z^* &= x_2 - ix_3 \end{aligned} \tag{3.1}$$

transforms equation (1.1) into the form

$$L[U] = U_{xx} + 4\, U_{zz^*} + F(x,z,z^*)\, U = 0. \tag{3.2}$$

Now let $U(z,z^*,x)$ be a regular solution of equation (3.2) and $V(z,z^*,x) (= V(z,z^*,x;\xi,\xi^*,\zeta))$ be a fundamental solution of $L[U] = 0$ with singularity at $(z,z^*,x) = (\xi,\xi^*,\zeta)$. Integrating the identity $VL[U] - UL[V]$ over the torus $D \times \Omega$ where Ω is the circle $|x-\zeta| = \delta > 0$ in the complex plane and D is the two dimensional cell as in figure 1 below

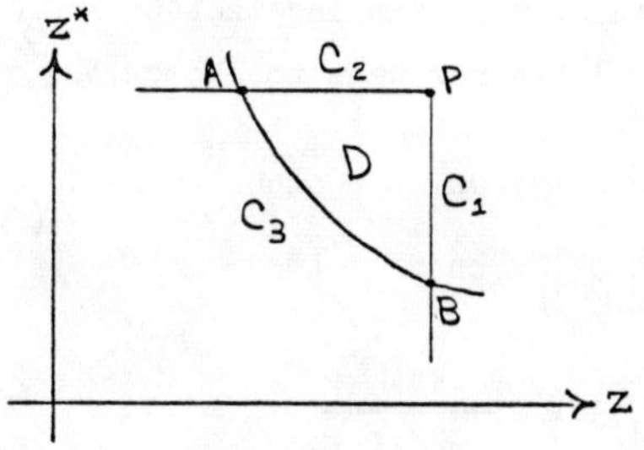

FIGURE 1.

gives ([3])

$$2\int_{\Omega}\Big[V(A,x)\,U(A,x) + V(B,x)\,U(B,x) - 2\,V(P,x)\,U(P,x)\Big]\,dx$$

$$+4\int_{C_1}\int_{\times\,\Omega} U\,V_{z^*}dz^*dx - 4\int_{C_2}\int_{\times\,\Omega} U\,V_z\,dzdx \qquad (3.3)$$

$$+\,2\int_{C_3}\int_{\times\Omega}(UV_{z^*} - VU_{z^*})dz^*dx - (UV_z - VU_z)\,dzdx = o\ ,$$

where we have used the fact that $dzdz^* = o$ on $\partial D \times \Omega$. Note that an expression of the form $V(A,x)$ is a function of three independent variables, i.e. $V(A,x) = V(z,z^*,x)$ $(= V(z,z^*,x;\xi,\xi^*,\zeta))$ where (z,z^*) are the coordinates of the point A in C^2. We now want to choose D and $V(z,z^*,x)$ such that equation (3.3) reduces to an integral representation of the solution to the Cauchy problem (1.1), (1.2). We first choose D such that C_3 is the intersection (in (z,z^*,x) space) of the plane $x = \zeta$ with the surface S and C_1 and C_2 are the intersections of this plane with the characteristic planes passing through the points A and P and B and P respectively where $P = (\xi,\xi^*,\zeta)$. In particular this implies that $C_3 = C_3(\zeta)$ is a (analytic) function of ζ. We now want to choose $V(z,z^*,x)$ such that

$$\begin{aligned} V_{z^*} &= o \qquad \text{on } z = \xi \\ V_z &= o \qquad \text{on } z^* = \xi^* \end{aligned} \qquad (3.4)$$

and

$$V(\xi,\xi^*,x) = \frac{1}{x-\zeta}\ . \qquad (3.5)$$

This will imply that in equation (3.3) the integrals along $C_1 \times \Omega$ and $C_2 \times \Omega$ vanish and that a residue arises at the point P from the integration around Ω. In order to construct such a function we recall (c.f. [8]) that $V(z,z^*x)$ is of the form

$$V = \frac{1}{R} + \sum_{\ell=1}^{\infty} U_\ell R^{2\ell-1} + W \tag{3.6}$$

where $R = \sqrt{(x-\zeta)^2 + (z-\xi)(z^* - \xi^*)}$, U_ℓ, $\ell = 1,2,\ldots,$ are analytic functions which can be computed recursively, and $W(z,z^*,x)(=W(z,z^*,x;\xi,\xi^*,\zeta))$ is a regular solution of $L[W] = 0$. Noting that $\frac{1}{R}$ satisfies the initial data (3.4), (3.5), it is seen that if we construct a regular solution W of $L[W] = 0$ satisfying the (entire) Goursat data

$$\begin{aligned} W &= -\sum_{\ell=1}^{\infty} U_\ell (x-\zeta)^{2\ell-1} \quad \text{on } z = \xi \\ W &= -\sum_{\ell=1}^{\infty} U_\ell (x-\zeta)^{2\ell-1} \quad \text{on } z^* = \xi^* , \end{aligned} \tag{3.7}$$

then the fundamental solution $V(z,z^*,x)$ defined by equation (3.6) satisfies the initial conditions (3.4), (3.5). The existence of such a function W is assured by Hormander's generalized Cauchy Kowalewski theorem ([11]). Note that due to the form of the initial conditions (3.4), (3.5) the function $V(z,z^*,x)$ is not uniquely determined. For example in the case when $f(\underset{\sim}{x}) = \lambda^2 =$ constant another choice for V is given by

$$V = \frac{\cos \lambda R}{R} \quad . \tag{3.8}$$

With the above choices of D and $V(z,z^*,x)$ equation (3.3) becomes

$$U(\xi,\xi^*,\zeta) = +\frac{1}{4\pi i}\int_\Omega \Big[V(A,x)U(A;x)+V(B,x)U(B,x)\Big]\,dx$$

$$+\frac{1}{4\pi i}\int_{C_3(\zeta)}\int_{\times\Omega} (UV_{z^*}-VU_{z^*})\,dz^*dx - (UV_z - VU_z)\,dzdx. \tag{3.9}$$

Equation (3.9) gives the desired integral representation of the solution to the elliptic Cauchy problem (1.1), (1.2) (in the case $n = 3$) for (ξ,ξ^*,ζ) sufficiently near the initial surface S and δ sufficiently large. By deforming the contour integrals in equation (3.9), and noting that $V(z,z^*,x)$ is an analytic function of x outside the branch cut between $\zeta+ i\sqrt{(z-\xi)(z^*-\xi^*)}$ and $\zeta-i\sqrt{(z-\xi)(z^*-\xi^*)}$, we can use the integral representation (3.9) to analytically continue $U(\xi,\xi^*,\zeta)$. Finally we note that in order for the second integral in equation (3.9) to be well defined we require that the Cauchy data for $U(z,z^*,x)$ be defined (for each fixed x) in a domain which is conformally symmetric with respect to $C_3(\zeta)$. The above analysis now implies the following theorem:

THEOREM: Let G be a neighborhood in the complex x plane containing the branch cut joining $\zeta\pm i\sqrt{(z-\xi)(z^*-\xi^*)}$ for all $(z,z^*)\in C_3(\zeta)$ and $(\xi,\xi^*)\in D$. Suppose furthermore that $C_3(\zeta)\times G$ lies in the domain of regularity of the Cauchy data for $U(z,z^*,x)$ and that for $x=\zeta$ this Cauchy data is analytic in a domain B which is conformally symmetric with respect to $C_3(\zeta)$. Let $B^* = \{z^*: \overline{z^*}\in B\}$. Then the restriction of $U(z,z^*,x)$ to the plane $x=\zeta$ is an analytic function of z and z^* in $D\cap B\times B^*$ and can be represented in the form of equation (3.9) with Ω replaced by G and the roles of (z,z^*,x) and (ξ,ξ^*,ζ) interchanged.

REFERENCES

1. D. COLTON, On the inverse scattering problem for axially symmetric solutions of the Helmholtz equation, Quart. J. Math. 22 (1971), 125-130.

2. D. COLTON, Cauchy's problem for almost linear elliptic equations in two independent variables, J. Approx. Theory 3 (1970), 66-71

3. D. COLTON, Improperly posed initial value problems for self-adjoint hyperbolic and elliptic equations, SIAM J. Math. Anal., to appear.

4. D. COLTON, The Analytic Theory of Partial Differential Equations, mimeographed lecture notes, University of Glasgow, 1972.

5. P.R. GARABEDIAN, Stability of Cauchy's problem in space for analytic systems of arbitrary type, J. Math. Mech. 9 (1960), 905-914.

6. P.R. GARABEDIAN, Partial differential equations with more than two independent variables in the complex domain, J. Math. Mech. 9 (1960), 241-271.

7. P.R. GARABEDIAN, Applications of the theory of partial differential equations to problems of fluid mechanics, in Modern Mathematics for the Engineer: Second Series, Edwin Beckenbach, editor, McGraw Hill, New York, 1961, 347-372.

8. P.R. GARABEDIAN, Partial Differential Equations, John Wiley, New York, 1964.

9. P. HENRICI, A survey of I.N. Vekua's theory of elliptic partial differential equations with analytic coefficieitns, Z. Angew. Math. Physics 8 (1957) 169-203.

10. C.D. HILL, Linear functionals and the Cauchy-Kowalewski theorem, J. Math. Mech. 19 (1969), 271-277.

11. L. HORMANDER, <u>Linear Partial Differential Operators</u>, Springer-Verlag, Berlin, 1964.

12. C. MÜLLER, Radiation patterns and radiation fields, <u>J. Rat. Mech. Anal</u>. 4 (1955), 235-246.

13. M. van DYKE, The supersonic blunt-body problem-review and extension, <u>J. Aero/Space Sciences</u> 25 (1958), 485-496.

14. V.H. WESTON, J.J. BOWMAN, and E. AR, On the inverse electromagnetic scattering problem, <u>Arch. Rat. Mech. Anal</u>. 31 (1968), 199-213.

Some properties of solutions of the Navier-Stokes equations

R.H. Dyer

In this talk I would like to focus attention on some growth properties and uniqueness results of classical, strongly differentiable, solutions of the Navier-Stokes equations. Naturally, the reason for speaking at this conference about these properties is simply that, in the main, they are obtainable by means of logarithmic convexity methods and provide an interesting application of such methods. In making my remarks I shall be assuming the existence of solutions of the Navier-Stokes equations in the various function classes that I shall subsequently specify, and to put my assumptions in context it may be helpful if I briefly recall some existence results associated with the initial value problem for the Navier—Stokes equations, a problem which has been the subject of intensive interest. For ease of exposition in a rather technical subject I state but a restricted version of it:

Let R^3 denote Euclidean 3-space, and let $\Omega \subset R^3$ be a nonempty bounded domain with boundary $\partial\Omega$. The problem is to find a velocity field $u(x,t)$ and pressure field $p(x,t)$ satisfying

$$\left.\begin{aligned} u_t + u.\text{grad}\, u &= -\text{grad}\, p + \Delta u \\ \text{div}\, u &= 0 \end{aligned}\right\}. \quad (x,t) \in \Omega \times (0,\infty)$$

and such that

$$u(x,0) = u_0(x) \quad \text{for all } x \in \bar{\Omega}$$

and

$$u(x,t) = 0 \quad \text{for} \quad (x,t) \in \partial\Omega \times [0,\infty),$$

where the divergence free vector field u_0 is prescribed. For simplicity I have omitted consideration of a body force term and have set the kinematic viscosity and density both to have unit value.

In physical terms we suppose the domain a rigid vessel filled with incompressible fluid; the fluid is initially set in motion and we are to determine its subsequent velocity distribution, subject to the Navier-Stokes equations and adherence to the boundary of Ω.

In connection with this problem a result of the following kind asserting the existence of a smooth solution for at least some finite interval of time is of substantial interest.

Theorem 1. (Kaniel and Shinbrot [14]).

Let $u_0 \in C^\infty(\bar{\Omega})$ and let $\partial\Omega$ be "sufficiently smooth" (see [14]). Then there exists $T > 0$ such that a $C^\infty(\bar{\Omega} \times [0,T])$ solution of the Navier-Stokes equations exists which assumes the initial data u_0 and zero boundary data on $\partial\Omega \times [0,T)$.

The proof of such a theorem is not a little complicated and demands a fair amount of preparation. Historically, it has been fashioned through the efforts of several mathematicians: Basic contributions concerning the existence of "weak" solutions of the initial value problem were made by Leray [18,19,20], Hopf [10], and Kiselev and Ladyzhenskaya [15], and subsequently, information about the regularity of weak solutions was obtained by Ito [11], Serrin [27], and Kaniel and Shinbrot [14]. I must refer you to their work for the detail and subtleties.

Concerning uniqueness of solutions of the initial value problem, results establishing uniqueness as a weak solution were obtained by Serrin [28]: provided that the boundary is smooth, the proof of uniqueness, forward in time, in the class of smooth solutions is an elementary matter. If Ω denotes a fixed bounded domain in R^3 with smooth boundary so that the application of Green's theorem is justified then one can readily establish

Theorem 2. (Serrin [26])

Let d be the diameter of Ω. Let $u^{(i)} \in C^2(\bar{\Omega} \times [0,T))$ and $p^{(i)} \in C^1(\bar{\Omega} \times [0,T))$ $(i = 1,2)$ be solutions of the Navier-Stokes equations such that $u^{(1)} = u^{(2)}$ on $\partial\Omega \times [0,T)$; let $-m$ (≤ 0) be a lower bound for the characteristic values of the deformation tensor $(D_{ij}) = (\frac{1}{2}(u_{i,j} + u_{j,i}))$ associated with $u^{(1)}$ in $[t_0,T)$; let $u = u^{(2)} - u^{(1)}$, and set $K = \frac{1}{2}\int_\Omega u^2\,dx$. Then

$$K(t) \leq K(t_0) \exp\,(2m - 6\pi^2/d^2)(t - t_0) \qquad 0 \leq t_0 \leq t < T$$

The proof of this result is based on the identity

$$\frac{d}{dt}K = -\int_\Omega \{(\text{grad } u : \text{grad } u) + u.D.u\}\,dx,$$

and use of the well known Poincaré inequality. From the result we can obtain information about stability, in a certain sense, and the knowledge that forward in time solutions of the initial value problem depend continuously (in the L^2 norm) upon the initial data. Moreover, the obvious corollaries below highlight other aspects of individual interest.

Corollary 1.

If $u^{(1)}(x,t_0) = u^{(2)}(x,t_0)$ for each $x \in \Omega$ then $u^{(1)} \equiv u^{(2)}$ for $(x,t) \in \Omega \times [t_0,T)$.

Corollary 2. (Kampé de Fériet [13])

If $u^{(1)} \equiv 0$ then $m = 0$ and so

$$K(t) \leq K(t_0) \exp\left(-\frac{6\pi^2}{d^2}(t - t_0)\right) \quad \text{for} \quad t \in [t_0,T).$$

Clearly, Corollary 1 affords a uniqueness theorem forward in time and Corollary 2 provides an exponential upper bound for the rate of decay of a solution.

Theorem 2 and its corollaries pose some interesting mathematical questions. In the first place is there backward in time, continuous (L^2) dependence upon the data at time t_0? Secondly, what is known about uniqueness backward in time? Thirdly, can lower bounds be obtained for the rate of decay of a solution and, furthermore, if the solution u is assumed to exist for all $t \geq 0$, what is known about the asymptotic behaviour of u as $t \longrightarrow +\infty$?

Concerning these questions let us begin by supplementing the result of Corollary 2 with one of the simpler versions of a theorem on lower bounds for the rate of decay. If $T = +\infty$, this theorem will also provide information about asymptotic behaviour, and in part, its inspiration was an attempt to improve an earlier result due to Edmunds [8] about such behaviour. I quote his result for comparison.

Theorem 3. (Edmunds [8])

Let $u \in C^2(\bar{\Omega} \times [0,\infty))$ be a classical solution of the Navier-Stokes equations such that $u(x,t) = 0$ if $(x,t) \in \partial\Omega \times [0,\infty)$. Suppose that as $t \longrightarrow +\infty$

$$\sup_{x\in\Omega} \left| \frac{\partial u_i}{\partial x_j}(x,t) \right| = O(e^{-\mu t})$$

for every $\mu > 0$ $(1 \leqslant i,j \leqslant 3)$. Then $u(x,t) = 0$ for $(x,t) \in \Omega \times [0,\infty)$.

In the statement and proof of the theorem on lower bounds we make use of the following notation:

Let Ω be a bounded domain in R^3 with smooth boundary $\partial\Omega$ so that Green's theorem is valid in Ω. For vector valued functions $g = (g_1, g_2, g_3)$ and $h = (h_1, h_2, h_3)$ in $C^1(\bar{\Omega} \times [0,T))$, $0 < T \leqslant +\infty$, set

$$(g(t),h(t)) = \int_\Omega g_i(x,t)h_i(x,t)dx \quad \text{and} \quad |g(t)|^2 = (g(t),g(t)) ,$$

so that $(.,.)$ denotes the inner product in $[L^2(\Omega)]^3$, and set

$$\|g(t)\|^2 = \int_\Omega |\text{grad } g(x,t)|^2 \, dx = \int_\Omega \frac{\partial g_j}{\partial x_i}(x,t) \frac{\partial g_j}{\partial x_i}(x,t)\, dx.$$

For brevity we also write $|g| = |g(t)|$: in referring to the Euclidean norm of the vector g we indicate its arguments,

$$|g(x,t)|^2 = g_i(x,t)\, g_i(x,t).$$

In the above notation, the promised theorem is

Theorem 4. (Ogawa [24])

Let $u \in C^2(\bar{\Omega} \times [0,T))$ and $p \in C^1(\bar{\Omega} \times [0,T))$ be a solution of the Navier-Stokes equations with $u = 0$ on $\partial\Omega \times [0,T)$. Suppose that $U(t) = \sup\{|u(x,t)| : x \in \Omega\} \in L^2([0,T))$.

(a) If $|u(t)| > 0$ for $0 \leqslant t_0 \leqslant t < T$, then

$$|u(t)| \geqslant |u(t_0)| \exp -\lambda(t-t_0) \text{ for } t_0 \leqslant t < T,$$

where λ is a positive constant depending on $u(t_0)$ and $\|U\|_{L^2}$.

[The hypothesis, $|u(t)| > 0$ in $[0,T)$ can be relaxed.]

(b) Let $T = +\infty$. If $|u(t)| = 0(\exp-\mu t)$ as $t \longrightarrow +\infty$ for each $\mu > 0$, then $u \equiv 0$.

Remark: The proof of this theorem is based on convexity type inequalities which are satisfied by solutions of certain differential inequalities in Hilbert space. To obtain a clear idea of the motive behind the operations undertaken in the proof below it may be helpful to consult the abstract papers on differential inequalities by Ogawa [23], Agmon and Nirenberg [2], and Cohen and Lees [3].

Proof: We observe first that

$$\begin{aligned}\frac{d}{dt}|u|^2 &= 2\,(u,u_t)\\ &= 2\,(u,\Delta u - u.\text{grad } u - \text{grad } p).\end{aligned}$$

Using Green's theorem, the fact that u has zero divergence, and the boundary conditions we find that

$$(u,u.\text{grad } u) = 0, \quad (u, \text{grad } p) = 0$$

and so

$$\frac{d}{dt}|u|^2 = -2\,\|u\|^2 . \tag{1}$$

Similarly, we find that

$$\frac{d}{dt}\|u\|^2 = -2(u_t, \Delta u)$$
$$= -2(u_t, u_t + u.\mathrm{grad}\ u)$$

Let $Q(t) = \|u(t)\|^2/|u(t)|^2$: If $t \in [t_0,T)$ then

$$|u|^4 \frac{d}{dt} Q = -2\{|u|^2(u_t, u_t + u.\mathrm{grad}\ u) - (u,u_t)(u,u_t + u.\mathrm{grad}\ u)\}$$
$$= -2\{|u|^2\ |u_t + \tfrac{1}{2} u.\ \mathrm{grad}\ u|^2 - \tfrac{1}{4}|u|^2\ |u.\ \mathrm{grad}\ u|^2 - (u,u_t + \tfrac{1}{2} u.\mathrm{grad}\ u)^2 + \tfrac{1}{4}(u,u.\mathrm{grad}\ u)^2\}\ .$$

Thus, using Schwarz inequality and the hypotheses of the theorem,

$$\frac{dQ}{dt} \leq \tfrac{1}{2}\frac{|u.\ \mathrm{grad}\ u|^2}{|u|^2} \leq \tfrac{1}{2} U^2 Q$$

and this is readily integrated to give

$$Q(t) \leq Q(t_0) \exp \tfrac{1}{2}\int_{t_0}^{t} U^2(s)\, ds\ , \quad t_0 \leq t < T. \tag{2}$$

From the identity (1) we find that

$$\frac{d}{dt}\log|u| = -Q\ .$$

Inserting (2) into this equality and integrating, we obtain for $t \in [t_0\ T)$

$$|u(t)| \geq |u(t_0)| \exp\{-Q(t_0)\int_{t_0}^{t}(\exp\int_{t_0}^{r}\tfrac{1}{2}U^2(s)\, ds)\, dr\}$$

Since, by hypothesis, $U \in L^2([0,T))$ we find

$$|u(t)| \geq |u(t_0)| \exp\{-\lambda(t-t_0)\},\quad t_0 \leq t < T$$

the desired inequality, where $\lambda = Q(t_0)\exp\tfrac{1}{2}\|U\|^2_{L^2}$ is a positive constant.

To relax the condition that $|u(t)| > 0$ for $0 \leq t_0 \leq t < T$ note that by Corollary 1 either $u \equiv 0$ in $[0,T)$ or $|u(0)| > 0$. Suppose that $|u(0)| > 0$ and also that there exists a $t \in [0,T)$ such that $|u(t)| = 0$. Clearly, in that event there exists a least positive value of t, say t^*, such that $|u(t^*)| = 0$. However, since the conclusions of part (a) of the theorem hold in $[0,t^*)$, by the continuity of $t \longmapsto |u(t)|$ they must hold at t^* also. In particular, $|u(t^*)| \geq |u(0)|\exp(-\lambda t^*) > 0$ which contradicts $|u(t^*)| = 0$. It follows then that either $u \equiv 0$ or $|u(t)| > 0$ in $[0,T)$.

Lastly, suppose the conclusion of part (b) is false. If $u \not\equiv 0$, then $|u(0)| > 0$ and from part (a) we obtain

$$e^{\lambda t}|u(t)| \geq |u(0)|, \quad 0 \leq t < +\infty.$$

But for this to be compatible with the hypothesis of (b) it must be that $|u(0)| = 0$ giving a contradiction. Thus $u \equiv 0$.

□

Now, rather similar results concerning the exponential decay of solutions, which satisfy zero Dirichlet boundary data, of certain parabolic partial differential equations with time independent coefficients have been obtained by Masuda [21]. Loosely speaking, he proves that for solutions that are defined for all real t and satisfy an exponential bound for all t, a maximum rate of decay as $t \longrightarrow +\infty$ can be found which depends on the exponential bound. In recent work, Ogawa [25], finds that a global theorem of this type is valid for parabolic differential inequalities in Hilbert space and that it is applicable to partial differential equations with time dependent coefficients. Assuming these results, it is tempting to conjecture that a similar theorem on the maximum rate of decay of solutions of the Navier-Stokes equations may be obtainable; such would be an interesting addition to the knowledge that, within a particular class, solutions of the Navier-Stokes equations are sandwiched between exponential bounds.

Let us now turn to the matter of backward uniqueness. As a basic result we have

Theorem 5 (Serrin [28])

Let Ω be a bounded domain in R^3 with smooth boundary $\partial\Omega$. Let $u^{(i)} \in C^2(\bar{\Omega} \times [0,T))$ and $p^{(i)} \in C^1(\bar{\Omega} \times [0,T))$ $(i = 1,2)$ be two solutions of the Navier-Stokes equations with $u^{(1)} = u^{(2)}$ on $\partial\Omega \times [0,T)$. If for some $t_0 \in (0,T)$, $u^{(1)}(x,t_0) = u^{(2)}(x,t_0)$ for all $x \in \Omega$, then $u^{(1)} \equiv u^{(2)}$.

Remark: Serrin's proof was patterned after techniques developed in a paper by Lees and Protter [17] on the unique continuation of solutions of parabolic differential equations and inequalities. However, a proof may perhaps more simply be obtained by using similar techniques to those adopted in establishing Theorem 4.

Proof (Ogawa [24])

Let $u = u^{(2)} - u^{(1)}$ and $p = p^{(2)} - p^{(1)}$; then u has to satisfy

$$u_t - \Delta u + u^{(1)}.\text{grad } u + u.\text{ grad } u^{(2)} = -\text{ grad } p$$

$$\text{div } u = 0$$

and u vanishes on the boundary. Just as in the proof of Theorem 4, we find that if $|u| > 0$ then

$$\frac{d}{dt} Q \leqslant \frac{1}{2} \; \frac{|u^{(1)}.\text{ grad } u + u.\text{ grad } u^{(2)}|^2}{|u|^2} \quad .$$

Setting

$$U_i(t) = \max\left\{\sup_\Omega |u^{(i)}(x,t)|, \sup_\Omega |\text{grad } u^{(i)}(x,t)|\right\},$$

we then have

$$\frac{d}{dt} Q \leqslant U_1^2 Q + U_2^2 \tag{3}$$

Also, it may easily be shown that

$$\frac{d}{dt} \log |u| = -Q - \frac{(u, u^{(1)}.\text{grad } u + u.\text{grad } u^{(2)})}{|u|^2}$$

from which it follows that

$$\frac{d}{dt} \log |u| \geqslant -Q - U_1 Q^{\frac{1}{2}} - U_2. \tag{4}$$

We are given that $u(.,t_0) \equiv 0$: Suppose that there exists $t_1 \in [0,t_0)$

such that $|u(t_1)| > 0$. Plainly, we may assume that $|u(t)| > 0$ for $t \in [t_1, t_0)$. Integrating (3) from t_1 to t, we find that $Q(t)$ is bounded by a constant depending on $u(t_1)$. From this fact together with (4) we obtain

$$|u(t)| \geq |u(t_1)| \exp\{- A (t - t_1)\}, \quad t \in [t_1, t_0)$$

where A is a constant depending on t_1. But then, because of the continuity of $t \longmapsto |u(t)|$, we conclude that $|u(t_0)| > 0$, contradicting the hypothesis that $u(.,t_0) = 0$. Thus $u \equiv 0$ on $[0, t_0)$.

Forward uniqueness follows from Corollary 1 of Theorem 2.

□

Finally in connection with the problems that I raised associated with Theorem 2, in the class of $C^2(\bar{\Omega} \times [0,T))$ solutions there is continuous dependence on the data at time t_0 forward in time. Using methods of logarithmic convexity, Knops and Payne [16] have determined a class of admissible solutions within which there is continuous dependence on data backward in time. Professor Knops outlined this result in his earlier talk at this conference.

So far my remarks have been confined to solutions of the Navier-Stokes equations in a bounded domain in R^3, and it is natural to scan 'the unbounded scene' for similar results. Concerning uniqueness, various theorems are known, but possibly the most interesting is that due to Graffi [9], the feature of his theorem being that it is proved without the strong assumptions about the convergence at infinity of the velocity field common in earlier papers on the topic. To be specific, if Ω is the complement of a bounded domain $\Omega_0 \subset R^3$ with smooth boundary $\partial\Omega$, Graffi considers solutions of the Navier-Stokes equations satisfying the following:

(i) u_i, $u_{i,t}$, $u_{i,j}$ $(i, j = 1,2,3)$ are continuous functions of (x,t) defined on $\Omega \times (-\infty,\infty)$ which for every finite time interval T are bounded in $\Omega \times T$ by a constant depending upon T. The second order derivatives of u_i are continuous in $\Omega \times (-\infty,\infty)$.

(ii) The pressure p has continuous first derivatives with respect to x_i in $\Omega \times (-\infty,\infty)$. As $r \longrightarrow +\infty$ $(r^2 = x_i x_i)$, p tends to a limit p_0 in such a way that

$p - p_0 = O(r^{-\frac{1}{2}-\epsilon})$, $\epsilon > 0$, uniformly in every finite time interval.

and he proves

Theorem 6 (Graffi [9])

Let $u^{(1)}$ and $u^{(2)}$ be solutions of the Navier-Stokes equations satisfying (i), (ii) (with the same p_0 in condition (ii) in both cases) and $u^{(1)}(x,t) = u^{(2)}(x,t)$ for all $(x,t) \in \partial\Omega \times (-\infty,\infty)$. If for some $t_0 \in (-\infty,\infty)$ $u^{(1)}(x,t_0) = u^{(2)}(x,t_0)$ for all $x \in \Omega$, then $u^{(1)}(x,t) = u^{(2)}(x,t)$ for all $(x,t) \in \Omega \times [t_0,\infty)$.

Actually, Graffi uses the condition $p - p_0 = O(r^{-1})$, and the improvement above is due to Serrin and is incorporated in the supplementing backward uniqueness theorem due to Edmunds [7], whose proof is modelled on Serrin's for the corresponding backward uniqueness theorem in the bounded case.

Theorem 7 (Edmunds [7])

Together with the hypothesis of Theorem 6, if $u = u^{(2)} - u^{(1)}$ and

$$\|u\|_\Omega = \left\{ \int_\Omega \frac{\partial u_i}{\partial x_j} \frac{\partial u_i}{\partial x_j} dx \right\}^{\frac{1}{2}}$$ exists for all $t \in (-\infty,\infty)$ then $u^{(1)} \equiv u^{(2)}$ for all $(x,t) \in \Omega \times (-\infty,\infty)$.

In passing, we note that a corresponding existence theorem is highly desirable.

The success of convexity methods in obtaining lower bounds and backward uniqueness in the case of bounded domains makes it natural to investigate by the same methods analogous postulates for the unbounded case. Under a variety of hypotheses, theorems concerning lower bounds and asymptotic behaviour are available, see [24] and [4]. However, concerning backward uniqueness, only a more restricted version of Theorem 7 has been found [24], extra spatial growth conditions being required.

For the remainder of this talk I would like to turn away from questions of evolution in time, and consider instead, how convexity methods may be used to derive results concerning the behaviour of solutions of the Navier-Stokes equations at a point. We shall consider first, a result pertaining to time independent solutions of the Navier-Stokes equations and it is convenient to preface its statement with two theorems, due to Agmon, which play a key role in its proof.

Let $B = \{x \in R^3 : |x| = (x_1^2 + x_2^2 + x_3^2)^{\frac{1}{2}} \leq r\}$ and let

$\Psi = \{v \in C^2(B) : v(x) = O(|x|^n)$ as $|x| \longrightarrow 0$ for all $n \in \mathbb{N}\}$

Theorem 8 (Agmon [1])

If $v \in \Psi$ and if there exists a constant $K > 0$ such that v satisfies the inequality

$$|\Delta v| \leq K\{|v| + |\text{grad } v|\}$$

throughout B, then $v \equiv 0$ in B.

The proof of this theorem, on the behaviour of a solution of an elliptic inequality at a point, is based upon an abstract theorem on differential inequalities in a Hilbert space, inequalities of the form

$$\| \ddot{u} - A(t)u \| \leq k(t) \{ \| \dot{u} \|^2 + (A(t)u,u)\}^{\frac{1}{2}} ,$$

where A(t) is a positive symmetric operator satisfying various conditions. For future needs, we note that Theorem 8 is also valid for vector-valued functions $v = (v_i)$ of x and, moreover, that B can have its centre at any point in R^3 provided that the conditions are modified appropriately.

Theorem 9 (Agmon [1])

If $v \in \Psi$ then the first derivatives of v each have a zero of infinite order at the origin.

For time independent (stationary) solutions of the Navier-Stokes equations one can prove directly, see [5], with the aid of Theorems 8 and 9 the following theorem,

Theorem 10.

Let Ω be a connected open set in R^3. Let $u \in C^3(\Omega)$ and $p \in C^2(\Omega)$ be a classical solution of the time independent Navier-Stokes equations. If at a point of Ω, taken to be the origin of coordinates, $u(x) = O(|x|^n)$ as $|x| \longrightarrow 0$ for every positive integer n, then u is identically zero throughout Ω.

Now, of course, this kind of unique continuation property is immediately deducible from the spatial analyticity of solutions of the Navier-Stokes equations in a wide class of circumstances: Concerning this, I refer you to the important contributions of Masuda [22] and Kahane [12]. Nonetheless, I think it of interest to see that such a unique continuation property can be obtained in an elementary manner, by appeal to the work of Agmon cited above. An outline of the proof of Theorem 10 will be given after a consideration of the development of a similar theorem for time-dependent solutions of the Navier-Stokes equations.

Concerning time dependent solutions it turns out that by using methods developed by Lees and Protter [17] one can obtain a sequence of theorems comparable with Theorems 8, 9 and 10:

$$\text{Let } \Phi = \{u \in C^3(B_T) : \lim_{|x| \to 0} e^{|x|^{-\beta}} |v| = 0 \text{ for every } \beta > 0,$$

$$\text{uniformly in } t \text{ for } t \in [0,T]\},$$

where $B_T = B \times [0,T]$.

Theorem 11 (Lees and Protter [17])

If $v \in \Phi$ and if there exists a constant $K > 0$ such that v satisfies the inequality

$$|\Delta v - v_t| \leq K\{|v| + |\text{grad } v|\}$$

throughout B_T, then $v \equiv 0$ in B_T.

Theorem 12.

If $v \in \Phi$, then $\lim_{|x|\to 0} e^{|x|^{-\beta}} |\partial v/\partial x_i| = 0$ $(i = 1,2,3)$ and

$\lim_{|x|\to 0} e^{|x|^{-\beta}} |\partial v/\partial t| = 0$ for every $\beta > 0$, uniformly in $[0,T]$.

Theorem 13.

Let $u \in C^4(\Omega \times [0,T])$ and $p \in C^2(\Omega \times [0,T])$ be a classical solution of the Navier-Stokes equations. If at some point 0 of Ω which we select to be the origin of coordinates $\lim_{|x|\to 0} e^{|x|^{-\beta}} |u| = 0$ for every $\beta > 0$ uniformly in t for $t \in [0,T]$, then u is identically zero throughout Ω.

Below, we outline the proof of Theorem 13; the proof of Theorem 10 from Theorems 8 and 9 exactly parallels that of Theorem 13 from Theorems 11 and 12.

Proof of Theorem 13 (an outline).

We use the equation for the convection of vorticity

$$\Delta\omega - \omega_t = u.\text{grad}\ \omega - \omega.\text{grad}\ u,$$

where $\omega = \text{curl}\ u$. Let B have centre at 0 and let its radius r be so chosen that B is contained in Ω: we show first that in B_T, $\omega \equiv 0$.

Because of the boundedness of u_i and $\partial u_i/\partial x_k$ in B_T $(i,k = 1,2,3)$, we have throughout B_T

$$|\Delta\omega - \omega_t| \leqslant K(|\omega| + |\text{grad}\ \omega|),$$

where K is a positive constant. By hypothesis $\omega \in C^3(B_T)$, and Theorem 12 ensures that the components ω_i $(i = 1,2,3)$ of ω belong to Φ. By the obvious generalisation of Theorem 11 for vector valued functions it then follows that $\omega \equiv 0$ in B_T.

Next, it can be shown that there exists $\phi \in \Phi$ such that $u = \text{grad}\ \phi$ and $\Delta\phi = 0$, from which it follows that $\phi \equiv 0$ in B_T and so $u \equiv 0$ in B_T.

Lastly, using the connectedness of Ω it can be shown that $u \equiv 0$ in $\Omega \times [0,T]$. □

Naturally, it is of interest to supplement the information contained in Theorems 10 and 13 about the behaviour of solutions at a point with information about the behaviour of solutions at infinity, and at least for stationary solutions of the Navier-Stokes equations some progress has been made in this direction making use of generalisations of Agmon's work, see [6].

References

1. S. Agmon, 'Unicité et convexité dans les problèmes différentiels' University of Montreal Press, (1966).

2. S. Agmon and L. Nirenberg, 'Properties of solutions of ordinary differential equations in Banach space', Comm. Pure Appl. Math. 16 (1963), 121-239.

3. P. J. Cohen and M. Lees, 'Asymptotic decay of solutions of differential inequalities', Pacific J. Math. 11 (1961), 1235-1249.

4. R. H. Dyer and D. E. Edmunds, 'Lower bounds for solutions of the Navier-Stokes equations', Proc. London Math. Soc. 18 (1968), 169-178.

5. R. H. Dyer and D. E. Edmunds, 'On the regularity of solutions of the Navier-Stokes equations', J. London Math. Soc. 44 (1969), 93-99.

6. R. H. Dyer and D. E. Edmunds, 'Asymptotic behaviour of solutions of the stationary Navier-Stokes equations', J. London Math. Soc. 44 (1969), 340-346.

7. D. E. Edmunds, 'On the uniqueness of viscous flows', Arch. Rational Mech. Anal. 14 (1963), 171-176.

8. D. E. Edmunds, 'Asymptotic behaviour of solutions of the Navier-Stokes equations', Arch. Rational Mech. Anal. 22 (1966), 15-21.

9. D. Graffi, 'Sul teorema di unicità nella dinamica dei fluidi', Annali di Mat. 50 (1960), 379-388.

10. E. Hopf, 'Über die Anfangswertaufgabe für die hydrodynamischen Grundgleichungen', Math. Nachrichten 4 (1951), 213-231.

11. S. Ito, 'The existence and uniqueness of regular solution of non stationary Navier-Stokes equation', J. Fac. Sci. Univ. Tokyo, Sec. I, 9 (1961), 103-140.

12. C. Kahane, 'On the spatial analyticity of solutions of the Navier-Stokes equations' Arch. Rational Mech. Anal. 33 (1969), 386-405.

13. J. Kampé de Fériet, 'Sur la décroissance de l'énergie cinétique d'un fluide visqueux incompressible occupant un domaine borné ayant pour frontière des parois solides fixes', Ann. Soc. Sci. Bruxelles 63 (1949), 36-45.

14. S. Kaniel and M. Shinbrot, 'Smoothness of weak solutions of the Navier-Stokes equations', Arch. Rational Mech. Anal. 24 (1967), 302-324.

15. A. A. Kiselev and O. A. Ladyzhenskaya, 'On the existence and uniqueness of the solution of the nonstationary problem for a viscous incompressible fluid', Izv. Akad. Nauk. SSSR 21 (1957), 655-680.

16. R. J. Knops and L. E. Payne, 'On the stability of solutions of the Navier-Stokes equations backward in time', Arch. Rational Mech. Anal. 29 (1968), 331-335.

17. M. Lees and M. H. Protter, 'Unique continuation for parabolic differential equations and inequalities', Duke Math. J. 28 (1961), 369-382.

18. J. Leray, 'Étude de diverses équations integrales non linéaires et de quelques problèmes que pose l'hydrodynamique', J. Math. Pures Appl. 12 (1933), 1-82.

19. J. Leray, 'Essai sur les mouvements plans d'un liquide visqueux que limitent des parois', J. Math. Pures Appl. 13 (1934), 331-418.

20. J. Leray, 'Sur le mouvement d'un liquide visqueux emplissant l'espace', Acta Math. 63 (1934), 193-248.

21. K. Masuda, 'On the exponential decay of solutions for some partial differential equations', J. Math. Soc. Japan 19 (1967), 82-90.

22. K. Masuda, 'On the analyticity and the unique continuation theorem for solutions of the Navier-Stokes equations', Proc. Japan Acad. 43 (1967), 827-832.

23. H. Ogawa, 'Lower bounds for solutions of differential inequalities in Hilbert space', Proc. Amer. Math. Soc. 16 (1965), 1241-1243.

24. H. Ogawa, 'On lower bounds and uniqueness for solutions of the Navier-Stokes equations', J. Math. Mech. 18 (1968), 445-452.

25. H. Ogawa, 'On the maximum rate of decay of solutions of parabolic differential inequalities', Arch. Rational Mech. Anal. 38 (1970), 173-177.

26. J. Serrin, 'On the stability of viscous fluid motions', Arch. Rational Mech. Anal. 3 (1959), 1-13.

27. J. Serrin, 'On the interior regularity of weak solutions of the Navier-Stokes equations', Arch. Rational Mech. Anal. 9 (1962), 187-195.

28. J. Serrin, 'The initial value problem for the Navier-Stokes equations', Proc. Symp. Non-linear Problems, Univ. of Wisconsin (1963), 69-98.

NON-UNIQUE CONTINUATION FOR CERTAIN ODE's IN HILBERT SPACE AND FOR UNIFORMLY PARABOLIC AND ELLIPTIC EQUATIONS IN SELF-ADJOINT DIVERGENCE FORM*

K. MILLER

I want to consider the problem of backward uniqueness for the uniformly parabolic equation:

$$u_t = \sum_{i,j=1}^{n} (a_{ij}(x,t)u_{x_j})_{x_i} \equiv \vec{\nabla}\cdot a\vec{\nabla}u \text{ in } \Omega\times[0,\infty) \,, \qquad \text{(a)}$$

(1)

$$\mathcal{a}\nabla u\cdot\nu = 0 \text{ on } \partial\Omega\times[0,\infty) \,, \qquad \text{(b)}$$

and the problem of unique continuation (and uniqueness for the Cauchy problem) for the uniformly elliptic equation:

$$\sum_{i,j=1}^{n} (a_{ij}(x)u_{x_j})_{x_i} \equiv \vec{\nabla}\cdot\mathcal{a}\vec{\nabla}u = 0 \text{ in } \Omega \,, \qquad (2)$$

where Ω is a bounded domain in R^n, $\vec{\nu}$ denotes the unit normal to $\partial\Omega$, and the symmetric coefficient matrix a has its eigenvalues in $[\alpha,\alpha^{-1}]$, with ellipticity constant $0 < \alpha < 1$. We construct examples of nonuniqueness for (1) when n = 2 and for (2) when n = 3; in each case α may be arbitrarily close to 1 and the coefficients are also Hölder continuous.

Equation (1a) or (2) are the equations of time dependent or steady state heat flow, with variable nonisotropic conductivity matrix $\mathcal{a}$, with constant heat capacity, and with the conductivity in each direction bounded above and

---

Talk given at the Symposium on Logarithmic Convexity and Non-Well Posed Problems, 21-24 March 1972, Heriot-Watt University, Edinburgh.

*Supported by a C.N.R. Visiting Professorship at Universita di Firenze and by N.S.F. grant

below. The "no flow" condition (1a) implies that $\partial\Omega$ is totally insulated.

Sometimes we will want to replace the Neumann boundary condition (1b) by the Dirichlet condition

$$u = 0 \text{ on } \partial\Omega\times[0,\infty) , \qquad (1b')$$

in which case we will speak of (1a) and (1b') combined as (1').

The famous results of De Giorgi [5] and Nash [7] established _a priori_ interior Hölder continuity of weak solutions of these equations with only measurable coefficients. This led to a tremendous development of the theory for second order elliptic and parabolic equations in divergence form, especially nonlinear equations. In some sense _a priori_ inequalities, those that are independent of the smoothness of the coefficients, show the fundamental elliptic (or parabolic) behaviour of the equation. I have been attempting to prove backward uniqueness (and a conjectured _a priori_ stability bound of logarithmic convexity type) for (1) for eight years.

Backward uniqueness for (1) with C^1 coefficients was shown by Lions and Malgrange [6] in 1960. However, probably the simplest proof is due to Agmon and Nirenberg [1] in 1964, and to Agmon [2] in 1966, who used the general method of logarithmic convexity. Carleman [4], long ago in 1939 established unique continuation for (2) with C^2 coefficients when $n = 2$. (It is possible that his proof can be carried through in this two-dimensional case with only measurable coefficients, but of this I am not certain.) For $n \geq 3$, unique continuation for (2) with $C^{2,1}$ coefficients was proved by Aronszajn [3] in 1957, and more simply with C^1 coefficients by Agmon [2] in 1966, using logarithmic convexity techniques once again. See [2] for references to other results by Cordes, Holmgren, Hörmander, Landis, Lees and Protter, and others.

An example of nonunique continuation was constructed by Pliś [9] for a uniformly elliptic equation in the nondivergence form:

$$\sum_{i,j=1}^{3} a_{ij}(x)u_{x_i x_j} + \sum_{i=1}^{3} b_i(x)u_{x_i} + c(x)u = 0 , \tag{3}$$

with Hölder continuous coefficients. Despite this example, the question of unique continuation for (2) has remained actively open. In the first place, it is well known that the solutions of the divergence equation (2) and the nondivergence equation (3) often exhibit extremely different behaviours when the coefficients become nonsmooth; in fact, the Pliś example occurs at exactly that degree of smoothness at which (3) can no longer be changed into the form (2) (plus bounded lower order coefficients). On the whole, solutions of equation (2) tend to behave like those of the Laplace equation, while solutions of (3) often exhibit quite striking pathology. In the second place, there is the strong physical interpretation of (1) and (2) in terms of heat flow (or diffusion, or electric flow). It seemed highly nonintuitive that an initially nonzero solution of (1) could manage to make itself vanish within a finite time T, or that a nontrivial solution of (2) could manage to wipe out all trace of itself on an open subset. Equation (3) on the other hand has no such physical interpretation.

Before describing the nonunicity examples for (1) and (2), I would like to mention a few aspects of my previous attempts to prove backward uniqueness and a priori stability for (1) and (1').

1. BACKWARD STABILITY WITH C^1 COEFFICIENTS FOR (1) AND (1')

The logarithmic convexity approach of Agmon and Nirenberg, when applied to the simple equation (1) (or (1')), can be redescribed so simply that it is worthwhile to give a sketch of it here. One writes (1) as an ODE in the Hilbert space $L^2(\Omega)$:

$$u' = - A(t)u, \ u(t) \varepsilon D_{A(t)}, \ t > 0 , \tag{4}$$

where A(t) is a symmetric operator with domain $D_{A(t)}$ consisting of all sufficiently smooth functions satisfying the boundary condition (1b) and with corresponding bilinear form:

$$a(t)(v,w) \equiv \int_\Omega \sum_{i,j} a_{ij} v_{x_j} w_{x_i} dx .$$

Now a(t)(v,w) is defined for all sufficiently smooth functions v and w, but integrating by parts we see that

$$\begin{aligned}(A(t)v,w) &= \int_\Omega \sum_{i,j} a_{ij} v_{x_j} w_{x_i} dx - \int_{\partial\Omega} w(\sum a_{ij} v_{x_j} \nu_i) d\sigma , \\ &= a(t)(v,w) \text{ if } v \varepsilon D_{A(t)} .\end{aligned}$$

Our first step is to differentiate $\log||u(t)||$ twice, as is usual in the logarithmic convexity method, and notice that a new term $- a'(u,u)/(u,u)$ appears, due to the change in the bilinear form for fixed u. Thus,

$$\Big[\log||u(t)||\Big]' = - \frac{(Au,u)}{(u,u)} = - \frac{a(u,u)}{(u,u)} ;$$

and hence,

$$\begin{aligned}\Big[\log||u(t)||\Big]'' &= - \frac{(u,u)[a(u',u)+a(u,u')+a'(u,u)]-a(u,u)[2(Au,u)]}{(u,u)^2} \\ &= \left[\frac{2(u,u)(Au,Au)-2(Au,u)^2}{(u,u)^2}\right] - \frac{a'(u,u)}{(u,u)}\end{aligned}$$

The first term is non-negative by the Cauchy inequality, as usual. Thus we have: <u>suppose the A(t) are nonincreasing (in the sense that a(t)(v,v) is nonincreasing for each fixed v), then $\log||u(t)||$ is convex for each sufficiently smooth solution of (4)</u>. The same result holds even more simply when (1b) is replaced by the Dirichlet condition (1b'), for then $D_{A(t)}$ is independent of t and u' will also be in $D_{A(t)}$ for sufficiently smooth solutions. I understand that similar results have been proved by Payne and Sather.

The second step is to change the variable on the t axis; i.e., let $d\tau/dt = \rho(t) > 0$ with $\tau(0) = 0$. With respect to the new variable $\tau(t)$ the equation (4) becomes

$$\frac{du}{d\tau} = -(\rho(t))^{-1}A(t)u \equiv B(\tau)u \,. \tag{5}$$

Now $B(\tau)$ has the bilinear form $\rho^{-1}a(v,v)$ and clearly we can make this nonincreasing by making ρ increase quickly enough. We must have

$$\left[\rho^{-1}a(v,v)\right]' = \frac{\rho a'(v,v)-\rho' a(v,v)}{\rho^2} \leqslant 0 \,,$$

or

$$\left[\log\rho(t)\right]' = \rho'/\rho \geqslant \frac{a'(v,v)}{a(v,v)} \,,$$

which latter term is bounded in terms of the ellipticity constant α and uniform bounds on the time derivatives a'_{ij} of the coefficients.

Thus by "distorting the t axis" by a sufficient amount we have forced $\log||u(t)||$ to be a convex function of the "distortion" $\tau(t)$, and we have the following stability bound for backward solutions of (4):

<u>if</u>

$$||u(T)|| \leqslant \varepsilon, \text{ and } ||u(0)|| \leqslant E \,, \tag{6}$$

<u>then</u>

$$||u(t)|| \leqslant \varepsilon^{\tau(t)/\tau(T)} E^{1-\tau(t)/\tau(T)}, \text{ for } 0 \leqslant t \leqslant T \,. \tag{7}$$

2. AN ABSTRACT EXAMPLE OF BACKWARD NONUNIQUENESS

Our example of backward nonuniqueness for (1) closely resembles an abstract example found in 1965 (but still unpublished) for an ODE in Hilbert space of the related form

$u' = -A(t)u$, $t > 0$, with (8a)

$0 < \alpha K \leqslant A(t) \leqslant \alpha^{-1}K$, where (8b)

K and A(t) are self-adjoint operators all having (8c)

the same domain.

Notice that equation (1') is in this form, with - K corresponding to the Laplacian operator Δ. Since the abstract example is simpler, but also shows certain limits to the logarithmic convexity approach, it is worthwhile to describe it here.

I had conjectured at one time that conditions (4) should be sufficient to imply that $\log||u(t)||$ is an "α^2-convex" function of t. We say that the decreasing function ϕ is <u>α^2-convex</u> on the interval I if for each triplet $t_1 < t_2 < t_3$ in I we have $(\phi(t_3)-\phi(t_2))/(t_3-t_2) \geq \alpha^{-2}(\phi(t_2)-\phi(t_1))/(t_2-t_1)$. The case $\alpha = 1$ of course corresponds to usual convexity. One may show, by a great variety of different proofs, that <u>this conjecture is true provided that the A(t) and K all commute with one another</u>. Without commutativity, however, <u>a priori</u> convexity of $\log||u(t)||$ is false, and in fact we now proceed to construct an example of backward nonuniqueness under conditions (8).

Notice that the graph of $\log||u(t)||$ has slope - (Au,u)/(u,u), which is comparable to - (Ku,u)/(u,u). The size of (Ku,u)/(u,u), however, is an indicator of whether u is mostly concentrated in low order eigenspaces of K, or in high order eigenspaces. If $\log||u(t)||$ is to manage to plunge to $-\infty$ within finite time T, the solution u(t) must somehow manage to rotate itself so as to concentrate in higher and higher order eigenspaces. The significance of logarithmic convexity for the case $A(t) \equiv K \equiv$ constant, however, is that this cannot then happen, for the high order eigencomponents of the solution then die out more quickly than the lower order eigencomponents; hence u(t) must concentrate more and more in the lower order eigenspaces and (Ku,u)/(u,u) must decrease.

Within the constraints of (8), however, it turns out that it is still possible for a solution to start within one eigenspace of K and rotate itself completely into a higher order eigenspace. Let us show that this is possible in the two dimensional case. We let

$$\bar{K} = \begin{pmatrix} 1 & 0 \\ 0 & \lambda \end{pmatrix}, \quad \bar{A}(t) = R^T(t) \begin{pmatrix} 1 & 0 \\ 0 & \lambda \end{pmatrix} R(t), \tag{9}$$

and let $\bar{u}(t)$ be the solution of

$$\bar{u}' = -\bar{A}(t)\bar{u}, \quad \bar{u}(0) = \begin{pmatrix} 1 \\ 0 \end{pmatrix}, \tag{10}$$

where $\lambda = \alpha^{-1}$ and $R(t)$ is a variable orthogonal matrix to be chosen appropriately. Notice that (8b) is always satisfied.

Suppose that in a first stage we choose $\bar{A}(t) = \bar{A}_1 \equiv R_1^T \bar{K} R_1$ where R_1 is constant. The solution of (10) is then

$$\bar{u}(t) = R_1^T \begin{pmatrix} e^{-t} & 0 \\ 0 & e^{-\lambda t} \end{pmatrix} R_1 \begin{pmatrix} 1 \\ 0 \end{pmatrix}. \tag{11}$$

This has a tendency to rotate the solution toward the low order eigenspace of A_1. By taking a sufficiently long duration t_1 for this stage (such that $e^{-t_1}/e^{-\lambda t_1} = 4$ will certainly do for example) and by choosing the principal axes of R_1 appropriately (see Figure 1) we can make $u(t)$ rotate exactly 45° toward the positive x_2 axis in this first stage. In a second stage of the same duration we can then take $A(t) \equiv A_2 \equiv R_2 \bar{K} R_2$ with R_2 chosen so as to rotate the solution through the final 45° toward the x_2 axis.

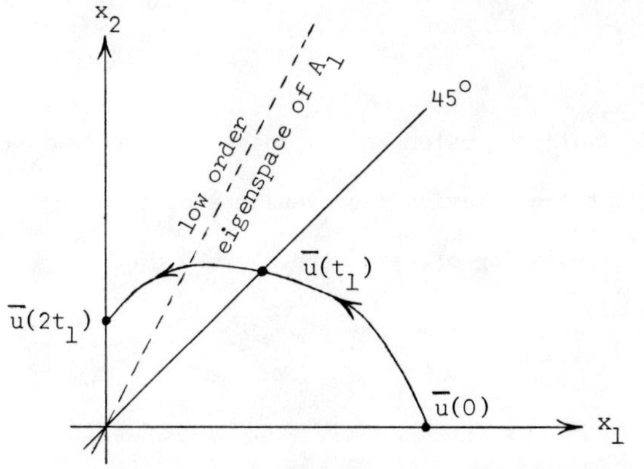

Figure 1. 90° rotation of $\bar{u}(t)$ in two stages

It should be clear that the 90° rotation just accomplished using a piecewise constant R(t) can also be accomplished using a smooth R(t) which begins and ends with I. One may, for example, choose a piecewise constant R(t) which rotates through 100°, then smooth the corners off in C^{∞} fashion such that the rotation is still $> 90^{\circ}$. Then we may vary a parameter (such as the duration of the rotating stages) and by continuity select that value of it which gives a rotation of exactly 90°. We state without further explanation: there exists a C^{∞} choice of R(t) in (9) such that $\bar{u}(t)$ in (10) rotates 90° from the x_1 axis into the x_2 axis in finite time T_o. In an initial phase R(t) is $\equiv$ I and $\bar{u}(t) = (e^{-t},0)^T$; in an intermediate phase R(t) varies and $\bar{u}(t)$ rotates through 90°; and in a final phase R(t) $\equiv$ I and $\bar{u}(t)$ is proportional to $(0,e^{-\lambda t})^T$. The magnitude of the solution decreases by a factor $r \equiv ||\bar{u}(T_o)||$ which can be made as small as we please merely by increasing the duration of the final phase.

We now patch together an ∞ of steps, each of which is essentially a carbon copy of the basic 2-dimensional example. We let K be the infinite matrix

$$K = \begin{pmatrix} 1 & & & & \\ & \lambda & & 0 & \\ & & \ddots & & \\ 0 & & & \lambda^n & \\ & & & & \ddots \end{pmatrix}, \tag{12}$$

considered as a self-adjoint operator on the space of l_2 column vectors $(x_o,x_1,\ldots)^T$. We adopt the notational convenience of starting time out anew with t = 0 at the beginning of each step.

The 0th step has duration T_o, initial value $u(0) = (1,0,\ldots)^T$, matrix

$$A(t) = \begin{pmatrix} R^T(t)\begin{pmatrix}1 & 0\\ 0 & \lambda\end{pmatrix}R(t) & & & 0\\ & \lambda^2 & & \\ & & \lambda^3 & \\ & & & \ddots \\ 0 & & & \end{pmatrix} = \begin{pmatrix} (\bar{A}(t)) & & 0\\ & \lambda^2 & \\ & & \ddots \\ 0 & & \end{pmatrix}, \tag{13}$$

and hence solution

$$u(t) = (\bar{u}_1(t), \bar{u}_2(t), 0, \ldots)^T \tag{14}$$

which rotates from the x_o axis completely into the x_1 axis. In general, the nth step has duration $T_n \equiv T_o/\lambda^n$, initial value $(0,\ldots,0,r^n,0,\ldots)^T$ on the x_n axis, and matrix

$$A(t) = \begin{pmatrix} 1 & & & & 0\\ & \ddots & & & \\ & & \lambda^{n-1} & & \\ & & & R^T(\lambda^n t)\begin{pmatrix}\lambda^n & 0\\ 0 & \lambda^{n+1}\end{pmatrix}R(\lambda^n t) & \\ & & & & \ddots \\ 0 & & & & \end{pmatrix} = \begin{pmatrix} 1 & & & & 0\\ & \ddots & & & \\ & & \lambda^{n-1} & & \\ & & & (\lambda^n \bar{A}(\lambda^n t)) & \\ & & & & \lambda^{n+2} \\ & & & & & \ddots \\ 0 & & & & \end{pmatrix}. \tag{15}$$

Hence, everything behaves here in the nth and (n+1)st components exactly like $\bar{u}(t)$, except λ^n times faster, and we must have the solution

$$u(t) = (0,\ldots,0,r^n\bar{u}_1(\lambda^n t), r^n\bar{u}_2(\lambda^n t), 0, \ldots)^T \tag{16}$$

which rotates from the x_n axis into the x_{n+1} axis. Figure 2 shows the graph of $\log||u(t)||$ for this example.

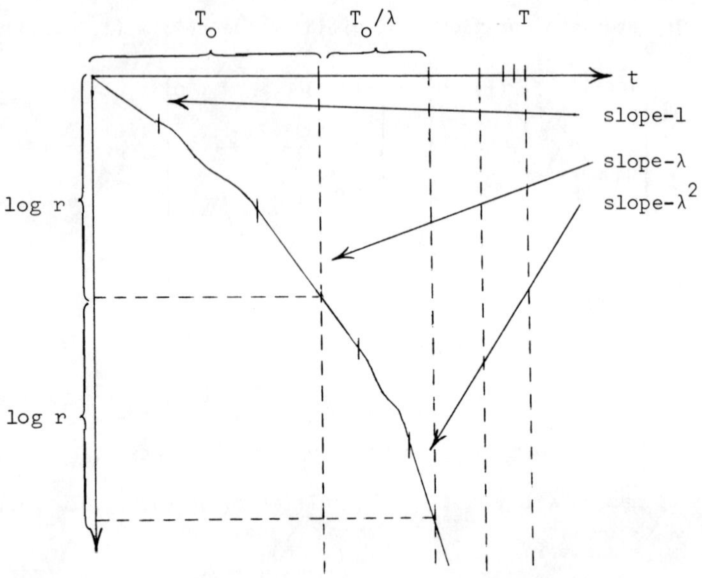

Figure 2. Graph of $\log||u(t)||$ for abstract example

Now $||u(t)|| \leqslant r^n$ in the nth interval, which $\to 0$ as $n \to \infty$ (or as $t \to T \equiv \sum T_n < \infty$). In general, the mth derivative has the bound $||u^{(m)}(t)|| \leqslant K_m r^n(\lambda^n)^m$ in the nth interval, where K_m is a bound for $||\bar{u}^{(m)}(t)||$. This $\to 0$ as $n \to \infty$ provided that r is taken sufficiently small that

$$r < \lambda^{-m} . \tag{17}$$

We then set

$$u(t) \equiv 0 \text{ and } A(t) \equiv K, \text{ for } t \geqslant T . \tag{18}$$

Since $T - t \sim \lambda^{-n}$ in the nth interval, condition (17) implies that $||u(t)|| = O(|T-t|^m)$ as $t \to T$. In this way we see that the strong derivatives $u^{(1)}, u^{(2)}, \ldots u^{(m)}$ exist also at $t = T$ and are zero there. We summarise:

Theorem 1. There exists an example of backward uniqueness for an ODE equation of form (4) on the Hilbert space l_2. The solution u(t) is $\equiv 0$ for $t \geqslant T$ but never 0 for $t < T$; u(t) is strongly C^m on $[0,\infty)$, where m can be made as large as we like. The operator A(t) is C^∞ (in any sense of the word) for $t \neq T$, but becomes discontinuous at $t = T$.

In the above example the final phase of the nth step has duration proportional to λ^{-n} and the magnitude $||u(t)||$ is reduced by the same factor r in each step; however, we can instead increase that duration, to n^{-2} say; thus the magnitude $||u(t)||$ is reduced by a factor $r_n < e^{-\lambda^{n+1}n^{-2}}$ in the nth step. In this way, <u>we may construct our example with u(t) strongly C^∞ on $[0,\infty)$</u>.

It is easy to see that there is enough leeway here to make u(t) and all its derivatives $\to 0$ without keeping the ratio between successive diagonal elements of K exactly a constant. In fact we may even let $K_{n+1,n+1}/K_{n,n} = \lambda_n$ where λ_n is a sequence of numbers which converge downward to 1, but not too quickly. Since the "ellipticity constant" α in the nth interval will equal λ_{n+1}^{-1}, in this way, <u>we may construct our example with a variable "ellipticity constant" $\alpha(t)$ in (8b) which tends to 1 as $t \to T$</u>.

Let Δ denote the unbounded self-adjoint operator on $L^2(0,\pi)$ corresponding to the Laplacian with zero Dirichlet conditions at the end-points. With respect to the basis sinx,sin2x,...,sinkx,... for L^2, $-\Delta$ corresponds to the infinite diagonal matrix

$$- \Delta \sim \begin{pmatrix} 1^2 & & & & 0 \\ & 2^2 & & & \\ & & \ddots & & \\ & & & k^2 & \\ 0 & & & & \ddots \end{pmatrix} . \tag{19}$$

We may now just pick a subsequence $k_1,k_2,\ldots,k_n,\ldots$ such that the ratio k_{n+1}^2/k_n^2 is always $\approx \lambda = \alpha^{-1}$. Then using our previous construction, but skipping over the intermediate eigenspaces, we can begin with a u(0) in the $\sin k_1 x$ eigenspace and cause it to rotate successively into the $\sin k_2 x,\ldots,$ $\sin k_n x,\ldots$ eigenspaces. In this way, <u>we may construct our example on $L^2(0,\pi)$ with K being the negative Laplacian $-\Delta$</u>.

It can be shown immediately, however, that the abstract operators $A(t)$ in the example just mentioned cannot be elliptic differential operators on $(0,\pi)$. It can be shown (due to variation diminishing properties [8] which follow directly from the maximum principle) that solutions of a parabolic differential equation on a rectangle (in one space dimension) with $u = 0$ on the lateral sides (or $u_x = 0$) cannot increase their number of changes of sign as time advances. Since our abstract solution begins as $\sin k_1 x$ (with $k_1 - 1$ changes of sign) and ends the first step proportional to $\sin k_2 x$ (with more changes of sign), it is impossible for a parabolic equation to copy the behaviour of even a single step of the abstract example (when $n = 1$).

All of the above results had been found by the summer of 1965. From that time until a short while ago I had hopes that the ellipticity of the operators in (1a) (i.e., the maximum principle) would somehow intervene and permit one to prove that $\log||u(t)||$ is α^2-convex for solutions of (1'). There is no time here to describe some of the attempts to prove this conjecture, or the tens of thousands of computer run tests for a discretised version without finding counter examples (all unfortunately with $n = 1$). (It still seems quite likely to me that α^2-convexity for (1) and (1') does hold when $n = 1$.) It now turns out, however, that variation diminishing properties fail drastically for $n > 1$.

3. STATEMENT OF NONUNICITY RESULTS FOR (1) AND (2)

Theorem 1. There exists an example of backward nonuniqueness on the halfspace $\Lambda = R^2\times[0,\infty)$ for a uniformly parabolic equation

$$u_t = ((1+A+a)u_x)_x + (bu_y)_x + (bu_x)_y + ((1+C+c)u_y)_y \text{ in } \Lambda . \tag{20}$$

(i) The solution $u(x,y,t)$ is C^∞ on Λ, $\equiv 0$ for $t \geq$ a certain positive value T, but never $\equiv 0$ in any open subset of $R^2\times[0,T)$.

(ii) The coefficients $a(x,y,t)$, $b(x,y,t)$, $c(x,y,t)$ are C^∞ on Λ and $\equiv 0$ for $t \geq T$.

(iii) The coefficients $A(t)$, $C(t)$ are Hölder continuous (of order 1/6) on $[0,\infty)$, C^∞ on $[0,T)$, and $\equiv 0$ for $t \geq T$.

(iv) All functions u, a, b, c are periodic in x and y with period 2π; u is symmetric about $x = j\pi$ and $y = j\pi$, j integer.

(v) Moreover, u satisfies the "no flow" condition (1a) on the sides $\partial\Omega\times[0,\infty)$ of the cylinder $\Omega\times[0,\infty)$, $\Omega = (0,\pi)\times(0,\pi)$, since both b and the normal derivative $\frac{\partial u}{\partial \nu}$ are $= 0$ there.

Theorem 2. There exists an example of nonunique continuation on the halfspace $\Lambda = R^2\times[0,\infty)$ for a uniformly elliptic equation

$$u_{tt} + ((1+A+a)u_x)_x + (bu_y)_x + (bu_x)_y + ((1+C+c)u_y)_y = 0 \text{ in } \Lambda . \tag{21}$$

Conditions (i)-(v) on u, a, b, c, A, C hold exactly as stated in Theorem 1.

4. OUTLINE OF THE PARABOLIC EXAMPLE

Our construction proceeds with an ∞ of steps of successively shorter duration $T_1, T_2, \ldots$ whose sum T is finite. Let us consider the nth step, in which the solution begins proportional to $\cos\lambda_n x\cdot e^{-\lambda_n^2 t}$, ends proportional to $\cos\lambda_{n+1}y\cdot e^{-\lambda_{n+1}^2 t}$, and at intermediate times is always a linear combination of $\cos\lambda_n x$ and $\cos\lambda_{n+1}y$. Here $\lambda_1, \lambda_2, \ldots$ is an increasing sequence of integers, to be chosen later. Each step will consist of three major phases (plus four "transition" phases between for purposes of smoothness only). These seven phases have durations λ_n^{-2}, λ_n^{-2}, λ_n^{-2}, $s_n\lambda_n^{-2}$, λ_n^{-2}, λ_n^{-2}, λ_n^{-2} respectively, where s_n, a certain sequence tending to ∞, is also to be chosen later. We then define $\mu_n = \lambda_{n+1}/\lambda_n$ and $\varepsilon_n^2 \equiv e^{-(\mu_n^2-1)s_n}$. We point out that μ_n will be $\tilde{\sim}$ 1 and ε_n will be $\tilde{\sim}$ 0, especially for large n.

Let η and β be two fixed C^∞ functions on $[0,1]$ with the following behaviour: $\eta(t)$ is $\equiv 0$ near $t = 0$, monotone on $[0,1]$, and $\equiv 1$ near $t = 1$;

$\beta(t)$ is $\equiv 0$ near $t = 0$ and $\equiv t - 1$ near $t = 1$. We now proceed to list the duration, solution, and coefficients (in the order $1+A+a$, b, $1+C+c$) for each phase. For notational convenience we start the time out anew with $t = 0$ at the beginning of each phase and also employ the notation "$\sim$", "is proportional to".

The "transition 1" phase has duration λ_n^{-2}, solution $u \sim \cos\lambda_n x \cdot e^{-\lambda_n^2 t}$, and coefficients 1, 0, $\psi(t)$, where $\psi(t) = 1 + (\mu_n^2 - 1)\eta(\lambda_n^2 t)$. This phase merely changes the coefficients smoothly from an initial 1, 0, 1 to a final 1, 0, μ_n^{-2}.

The "seed" phase has duration λ_n^{-2}, solution $u \sim \cos\lambda_n x \cdot e^{-\lambda_n^2 t} + \eta(\lambda_n^2 t) c_n \cos\lambda_{n+1} y \cdot e^{-\lambda_n^2 t}$, and coefficients $1 + a$, b, μ_n^{-2}, where the rather complicated $a(x,y,t)$ and $b(x,y,t)$ will be described later. This phase introduces a tiny $\cos\lambda_{n+1} y$ component into the solution.

The "transition 2" phase has duration λ_n^{-2}, solution $u \sim \cos\lambda_n x \cdot e^{-\phi(\lambda_n^2 t)} + \varepsilon_n \cos\lambda_{n+1} y \cdot e^{-\lambda_n^2 t}$, and coefficients $\phi'(\lambda_n^2 t)$, 0, μ_n^{-2}, where $\phi(t) = t + (\mu_n^2 - 1)\beta(t)$. This phase smoothly changes the decay rate of the first component from an $e^{-\lambda_n^2 t}$ rate to an $e^{-\lambda_{n+1}^2 t}$ rate. It does not change the 1 to ε_n ratio of the two components.

The "distorted decay" phase has duration $\lambda_n^{-2} s_n$, solution $u \sim \cos\lambda_n x \cdot e^{-\lambda_{n+1}^2 t} + \varepsilon_n \cos\lambda_{n+1} y \cdot e^{-\lambda_n^2 t}$, and coefficients μ_n^2, 0, μ_n^{-2}. Since $\varepsilon_n^2 = e^{-(\lambda_{n+1}^2 - \lambda_n^2)\lambda_n^{-2} s_n}$, this phase reverses the initial 1 to ε_n ratio of the two components to a final ε_n to 1 ratio.

The "transition 3" phase has duration λ_n^{-2}, solution $u \sim \varepsilon_n \cos\lambda_n x \cdot e^{-\lambda_{n+1}^2 t} + \cos\lambda_{n+1} y \cdot e^{-\phi(\lambda_n^2 t)}$, and coefficients μ_n^2, 0, $\mu_n^{-2}\phi'(\lambda_n^2 t)$, where $\phi(t) = \mu_n^2 t + (\mu_n^2 - 1)\beta(1-t)$. This phase smoothly changes the decay rate of both components to the same $e^{-\lambda_{n+1}^2 t}$ rate. It does not change the ε_n to 1 ratio of the two components.

The "removal phase" has duration λ_n^{-2}, solution $u \sim \eta(\lambda_n^2(1-t))\varepsilon_n \cos\lambda_n x \cdot e^{-\lambda_{n+1}^2 t} + \cos\lambda_{n+1} y \cdot e^{-\lambda_{n+1}^2 t}$ and coefficients μ_n^2, b, 1 + c, where the b(x,y,t) and c(x,y,t) will be described later. This phase is quite similar in all respects to the previous "seed" phase, except that it removes a tiny component from the solution.

The "transition 4" phase has duration λ_n^{-2}, solution $u \sim \cos\lambda_{n+1} y \cdot e^{-\lambda_{n+1}^2 t}$, and coefficients $\psi(\lambda_n^2 t)$, 0, 1, where $\psi(t) = \mu_n^2 + (1+\mu_n^2)\eta(t)$. This phase merely changes the coefficients smoothly from μ_n^2, 0, 1 to 1, 0, 1.

5. DERIVATION OF THE a, b

It is convenient to normalise the geometry, expanding the x, y and t scales by factors of λ_n, λ_{n+1}, and λ_n^2 respectively. That is, we consider $\tilde{u}$ defined by

$$\tilde{u}(x,y,t) = u(x/\lambda_n, y/\lambda_{n+1}, t/\lambda_n^2) \sim \cos x \cdot e^{-t} + \eta(t)\varepsilon_n \cos y \cdot e^{-t} , \tag{22}$$

which must be the solution of an equation with coefficients $1 + \tilde{a}$, $\tilde{b}$, 1.

Since $\cos x \cdot e^{-t}$ and $\cos y \cdot e^{-t}$ are already solutions of the equation with coefficients 1, 0, 1, the perturbations $\tilde{a}$ and $\tilde{b}$ need only take care of the $\eta'(t)$ term in the equation and we are led to the following <u>perturbation equation</u>:

$$(\tilde{a}\sin x)_x = - \eta\varepsilon_n \sin y \tilde{b}_x - \sin x \tilde{b}_y - \eta'\varepsilon_n \cos y . \tag{23}$$

By considering $\tilde{b}$ of the form f(y) s(x,y), where s is a certain solution of the first order PDE

$$\eta\varepsilon_n \sin y \frac{\partial s}{\partial x} + \sin x \frac{\partial s}{\partial y} = 0 , \tag{24}$$

one can construct (for $\varepsilon_n \lesssim$ a certain $\varepsilon_o > 0$) a $\tilde{b}$ which is = 0 on $x = j\pi$ and $y = j\pi$ as desired and such that the right hand side of (23) has mean value zero across horizontal lines between $x = j\pi$ and $x = (j+1)\pi$. This $\tilde{b}$ inserted in (23) then determines an a as desired. Moreover, one may show that each derivative of $\tilde{a}(x,y,t)$ and $\tilde{b}(x,y,t)$ is bounded by a constant (depending on the order of differentiation) times ε_n.

Construction of the coefficients b and c for the "removal" phase is completely analogous.

6. PUTTING THE PIECES TOGETHER

One now stacks the solutions and coefficients for the various phases and steps "end to end", after first multiplying the formulae ($u \sim \cos\lambda_n x \cdot e^{-\lambda_n^2 t}$ etc.) in each phase by an appropriate magnitude constant in order to maintain continuity. Notice that the magnitude of u in each phase of the nth step is decreasing at least at an $e^{-\lambda_n^2 t}$ rate.

Now with the proper choice of λ_n and s_n we can make $T \equiv \sum T_n = \sum(6+s_n)\lambda_n^{-2}$ finite, u and the a, b, c $\to 0$ (in C^∞ fashion) and the A(t), C(t) $\to 0$ (in Hölder continuous fashion) as $t \to T$. It suffices for example to choose $\lambda_n = (n+N)^3$, $s_n = (n+N)^4$, where the integer N is taken sufficiently large to keep ε_n, a, b, c, A, C small also during the initial steps. One then sets $u \equiv a \equiv b \equiv c \equiv A \equiv C \equiv 0$ for $t \geq T$.

7. THE ELLIPTIC EXAMPLE

The elliptic construction is extremely similar to the parabolic case. The solution in the nth step begins proportional to $\cos\lambda_n x \cdot e^{-\lambda_n t}$ and ends proportional to $\cos\lambda_{n+1} y \cdot e^{-\lambda_{n+1} t}$. The perturbation equation for $\tilde{a}$ and $\tilde{b}$ in the "seed" phase turns out to be essentially in the same form as (23) and the "transition 2" and "transition 3" phases require a bit more attention.

REFERENCES

[1] Agmon, S., Unicité et convexité dans les problèmes différentiels, Seminar Univ. di Montréal, Les Presses de l'Univer. Montréal, 1966.

[2] Agmon, S., and Nirenberg, L., Properties of solutions of ordinary differential equations in Banach space, Comm. Pure and Applied Math., 16 (1963), pp. 121-239.

[3] Aronszajn, N., A unique continuation theorem for solutions of elliptic partial differential equations or inequalities of second order, J. Math. Pures Appl., 36 (1957), pp. 235-249.

[4] Carleman, T., Sur un problème d'unicité pour les systèmes d'équations aux dérivées partielles à deux variables indépendentes, Ark. Mat. Astr. Fys. 26B (1939) No. 17, pp. 1-9.

[5] De Giorgi, E., Sulla differenziabilità e l'analiticità delle extremali degli integrali multipli, Mem. Acc. Sc. Torino, III, 3 (1957), pp. 25-43.

[6] Lions, J., and Malgrange, B., Sur l'unicité rétrograde, Math. Scand., 8 (1960), pp. 277-286.

[7] Nash, J., Continuity of the solutions of parabolic and elliptic equations, Amer. J. Math., 80 (1958), pp. 931-954.

[8] Nickel, K., Gestaltaussagen über Lösungen parabolischer Differentialgleichungen, J. Reine Angew. Math., 211 (1962), pp. 78-94.

[9] Pliś, A., On nonuniqueness in the Cauchy problem for an elliptic second order differential equation, Bull. Acad. Polon. Sci., Ser. Sci. Math. Astron. Phys., 11 (1963), pp. 95-100.

Department of Mathematics

University of California

Berkeley, California

LOGARITHMIC CONVEXITY AND THE CAUCHY PROBLEM FOR $P(t)u_{tt} + M(t)u_t + N(t)u = 0$ IN HILBERT SPACE*

Howard A. Levine**

I. INTRODUCTION

Let H be a Hilbert space, real or complex, with inner product $(,)$ and norm $\| \; \| = \sqrt{(,)}$. Let $T > 0$ or $T = +\infty$ and for $t \in [0,T)$, let $D(t) \subseteq H$ be a dense linear subspace of H. Let $M(t)$ and $N(t)$ be linear operators (bounded or unbounded) mapping $D(t)$ into H. In [9] and [10], questions of uniqueness and stability for the following (generally nonwell posed) problems were discussed.

Problem I. $M(t)\frac{du}{dt} = N(t)u + f_1(t,u)$; $t \in [0,T)$

$u(0)$ prescribed.

Problem II. $M(t)\frac{d^2u}{dt^2} = N(t)u + f_2(t,u,du/dt)$; $t \in [0,T)$

$u(0)$ and $du/dt(0)$ prescribed.

Here $u : [0,T) \longrightarrow H$ is such that the required number of strong derivatives are $D(t)$ valued. The "nonlinear" terms f_1 and f_2 were assumed to satisfy, for $u_1, u_2 : [0,T) \longrightarrow D(t)$, inequalities of the form

---

*This research was supported in part by N.S.F. Contract GP 7041X at the University of Minnesota and in part by the Battelle Institute, Advanced Studies Center of Geneva, Switzerland. This support is gratefully acknowledged.

**Permanent address: School of Mathematics, Institute of Technology, University of Minnesota, Minneapolis, Minnesota 55455.

(1) $$\|f_1(t,u_1) - f_1(t,u_2)\| \leqslant k_1\alpha(t,w)$$

or

(2) $$\|f_2(t,u_1,du_1/dt) - f_2(t,u_2,du_2/dt)\|$$
$$\leqslant k_1\alpha(t,w) + k_2\alpha(t,dw/dt)$$

where $w = u_1 - u_2$ and

(3) $$\alpha^2(t,w) \equiv |(w(t),M(t)w(t))| + \mu\int_0^t |(w(\eta),M(\eta)w(\eta))|d\eta$$

and where k_1, k_2 and μ are nonnegative locally bounded, measurable functions of t. In these papers a number of examples were cited from the applications. There, as well as in the applications found in [5], [6], [12], [13] and [14], for example, it was not generally assumed that $N(t)$ (or at least its symmetric part) was definite, thus leading to the (generally) ill posed nature of these problems.

In [1] and [2] the authors were concerned with abstract differential inequalities of the form

(α) $$\left\|\frac{du}{dt} - B(t)u\right\| \leqslant \Phi(t)\{\|u\|^2 + \int_t^T \omega(\eta)\|u\|^2 d\eta\}^{\frac{1}{2}}$$

(β) $$\left\|\frac{d^2u}{dt^2} - B(t)u\right\| \leqslant \Phi(t)\{\left\|\frac{du}{dt}\right\|^2 + (u,Bu)\}^{\frac{1}{2}}$$

and

$$(\delta) \qquad \left\| \frac{d^2u}{dt^2} + 2L(t)\frac{du}{dt} + B(t)u \right\|$$

$$\leqslant \gamma(t)\{\|\frac{du}{dt}\|^2 + Q_t(u,u)\}^{\frac{1}{2}} + \gamma_o^2\|u\|$$

where Φ, ω, γ, γ_o are nonnegative, scalar functions of $t \in [0,T)$. Although in [1] and [2] the authors were primarily interested in questions of unique continuation for solutions to elliptic equations, their results also yield uniqueness and stability theorems for certain initial-boundary value problems in partial differential equations. (Note that questions of uniqueness and stability for solutions to Problems I and II can be reduced to the corresponding questions for the null solution to the corresponding abstract evolutionary differential inequalities provided the upper bound on the right hand side is of the form given by the right hand side of (1) or (2)).

In [9], Problem I was studied and the results compared to those for inequality (α). Likewise, in [10], Problem II was examined, sufficient conditions on $M(t)$ and $N(t)$ given in order to insure uniqueness and stability and the results then compared with those for (β) in [1]. It is the purpose of this paper to examine

Problem III. $P(t)\frac{d^2u}{dt^2} + M(t)\frac{du}{dt} + N(t)u = f(t,u,du/dt), \quad t \in [0,T)$

$u(0)$ and $du/dt(0)$ prescribed.

This paper is divided into four sections. In the first section we show how Problem III can sometimes be reduced to Problem I or Problem II. In the second section we treat the linear version of Problem III ($f \equiv 0$) in some detail, giving sufficient, but fairly general, conditions on the operator families $P(\cdot)$, $M(\cdot)$ and $N(\cdot)$ in order to insure uniqueness of solutions and continuous dependence on the initial data as well as continuous dependence on the value of the operator; $\mathcal{L}(t)u \equiv Pu_{tt} + Mu_t + Nu$. There we also indicate how the corresponding results can be obtained for the nonlinear version of Problem III. In the third section we mention some applications of the theory to the theory of partial differential equations citing some interesting examples from the literature. We also give two extensions of the results of the second section which are needed to prove uniqueness in certain problems to which the results of Section II do not directly apply. Finally, in the fourth section we compare our theorems to those of Agmon [1] and Lions [11].*

We wish to emphasise here that while we are only concerned with Problem III from an abstract point of view in this paper, we have in mind application of these results to initial and initial-boundary value problems for equations and systems of equations where the coefficients of the partial differential operator may not be analytic (in fact C^1 smoothness is all that we shall ever require on the principle part of the operator by way of regularity); where the coefficients of the principle part do not satisfy any conditions with respect to "type" so that the equation (5) may be of mixed type or even of no type. In equations with non-constant coefficients of the form in the title very little is known with regard to uniqueness and stability (and even less with regard to existence) if all the operators do not satisfy so called coersivity conditions. See Lions [11], for example.

* We emphasise that we are concerned only with the uniqueness question here. As Lions [11] remarks, the hypotheses needed for existence in such problems are usually different from those needed for uniqueness.

In a future work we hope to give some wider applications of the results presented herein. However, we shall give a few simple, but non-trivial, examples of Problem III to which our results can be applied and which do occur in the applications.

II. SOME SPECIAL CASES

In this section we shall assume that f satisfies the following Lipschitz condition. (Here $u_{,t} \equiv du/dt$ in the strong sense):

$$(4) \qquad ||f(t,u_1,u_{1,t}) - f(t,u_2,u_{2,t})|| \leq \lambda_1\alpha(t,w) + \lambda_2\alpha(t,w_{,t})$$

where λ_1, λ_2 and μ (below) are nonnegative, locally bounded functions defined on $[0,T)$, where

$$(5) \qquad \alpha(t,w) \equiv |(w(t),P(t)w(t))| + \mu\int_0^t |(w(\eta),P(\eta)w(\eta))|d\eta$$

and $w = u_1 - u_2$. We impose the following hypotheses, valid for each $t\varepsilon[0,T)$, on the operators $P(t)$, $N(t)$ and $M(t)$;

P-I. $P(t)$ is symmetric and there is a constant $\lambda \geq 0$ (independent of $t\varepsilon[0,T)$) such that $(P(t)x,x) \geq \lambda(x,x)$ for all $x\varepsilon D(t)$.

M-I. There is a nonnegative measurable function $\gamma\varepsilon\mathcal{L}^\infty_{loc}([0,T))$ such that for all $t\varepsilon[0,T)$ and $x\varepsilon D(t)$,

$$||M(t)x||^2 \leq \gamma(t)(x,P(t)x) .$$

These are all the conditions needed to reduce Problem III to Problem II. One simply defines

$$(6) \qquad f_1(t,u,u_{,t}) \equiv f(t,u,u_{,t}) - M(t)u .$$

Then one obtains, upon application of the triangle inequality

$$(7) \qquad ||f_1(t,u_1,u_{1,t}) - f_1(t,u_2,u_{2,t})|| \leq ||f^*|| + ||Mw||$$

where $w = u_1 - u_2$ and $f^* = f(t,u_1,u_{1,t}) - f(t,u_2,u_{2,t})$.

Squaring both sides of (7), using the inequality $2|ab| \leq a^2 + b^2$ on the cross term on the right and then M-I and P-I and once more taking square roots , we see that there follows an upper bound of the same form as on the right hand side of (2) with M replaced by P .

The question of uniqueness of solutions to Problem III can be reduced to the corresponding question for Problem I under the assumption of (4) and (5) as follows : Let P-I hold and suppose

N-I . The hypothesis M-I holds for $N(\cdot)$.

Assume in addition that

P-II . $(u,P(t)u)$ is differentiable for all $u \in C^2([0,T),H)$ and there is a nonnegative function $k(t) \in \mathcal{L}^\infty_{loc}([0,T))$ such that

$$\left|\frac{d}{dt}(u,P(t)u) - 2\mathrm{Re}(u,P(t)u_t)\right| \leq k(t)(u_1P(t)u) .$$

Suppose, moreover, that u_1 and u_2 are solutions of Problem III which $u_1(0) = u_2(0)$ and let $w = u_1 - u_2$. Let

(8) $$f_1(t,u,du/dt) = f(t,u,du/dt) - N(t)u = P(t)u_{tt} + M(t)u_t .$$

Let $f_1^i = f_1(t,u_i,du_i/dt), i = 1,2$ and f^* be as before, then we obtain

$$\|f_1^1 - f_1^2\| \leq \|f^*\| + \|N(t)w\| .$$

Whence it follows from N-I and (4), (5) that for suitable γ_1, γ_2

$$\|Pw_{tt} + Mw_t\| \leq \gamma_1 \alpha(t,w) + \gamma_2 \alpha(t,w_t) \tag{9}$$

where $\alpha(t,w)$ is given by (5) . Now let

$$v(t) = \frac{dw}{dt}(t) \tag{10}$$

so that

$$w(t) = \int_0^t v(\eta)d\eta \qquad (w(0) = 0)$$

where the integral is taken in the strong sense. Then (9) becomes

$$\|Pv_t + Mv\| \leq \gamma_1 \alpha(t,w) + \gamma_2 \alpha(t,v) \ . \tag{11}$$

The reduction to the desired differential inequality will be complete if we can show that there is an appropriate $\gamma(t) \geq 0$, $\gamma \in \mathcal{L}^\infty_{loc}$, such that

$$\alpha(t,w) \leq \gamma(t)\alpha(t,v) \tag{12}$$

To do this, in view of P-I and the definition of α , it suffices only to show that

$$(w,P(t)w) \leq \gamma(t) \int_0^t (v(\eta),P(\eta)v(\eta))d\eta \tag{13}$$

For then,

$$\int_0^t (w,Pw)d\eta \leq t\cdot \operatorname*{ess.sup}_{0\leq s<t} \gamma(s) \int_0^t (v(\eta),P(\eta)v(\eta))d\eta \ .$$

Now let

$$g(t) = (w(t),P(t)w(t))$$

Then,

$$\begin{aligned} g'(t) &= 2\mathrm{Re}(w_t,Pw) + \frac{d}{dt}(w,Pw) - 2\mathrm{Re}(w_t,Pw) \\ &\leq 2(w_t,Pw_t)^{\frac{1}{2}}(w,Pw)^{\frac{1}{2}} + k(t)(w,Pw) \end{aligned}$$

Thus,

$$g'(t) \leq (v,Pv) + (1+k(t))g(t) \ . \tag{14}$$

Denoting $1+k$ by μ , we have

$$\frac{d}{dt}\left[g(t)\exp(-\int_0^t \mu(\eta)d\eta)\right] \leq (v,Pv)\exp(-\int_0^t \mu(\eta)d\eta) \ .$$

Since $g(0) = 0,(w(0) = 0)$, we see that, after a quadrature,

$$\begin{aligned} g(t) &\leq \int_0^t (v,P(\eta)v)\exp(\int_\eta^t \mu(s)ds)d\eta \\ &\leq (\exp\int_0^t \mu(\eta)d\eta) \cdot \int_0^t (v(\eta),P(\eta)v(\eta))d\eta \ . \end{aligned} \tag{15}$$

Clearly (15) is of the same form as (13).

Consider for the moment the linear version of Problem III with $P=I$. The equation is

$$u_{tt} + Mu_t + Nu = 0 \qquad \left(u_t \equiv \frac{du}{dt}\right) . \tag{16}$$

As in ordinary differential equations, we get the equivalent system

$$\begin{cases} u_t = v \\ v_t = -Mv - Nu . \end{cases} \tag{17}$$

Letting

$$U = \begin{pmatrix} u \\ v \end{pmatrix} \varepsilon\ C^1([0,T);H\times H),$$

we have

$$U_t = \mathcal{N} U \tag{18}$$

where

$$\mathcal{N} = \begin{pmatrix} 0 & I \\ -N & -M \end{pmatrix} .$$

For the scalar product on $H\times H$ we take

$$[U_1,U_2] = (u_1,u_2) + (v_1,v_2)$$

where $U_i = \begin{pmatrix} u_i \\ v_i \end{pmatrix}$ (i = 1,2). Denote by $|||\ \ |||$ the corresponding norm. Assume for the moment that M and N and consequently η are time independent. Then (18) becomes

$$U_t = \eta_+ U + \eta_- U \tag{19}$$

where

$$\eta_+ \equiv \tfrac{1}{2}\begin{pmatrix} 0 & I-N \\ I-N & 2M \end{pmatrix}$$

and

$$\eta_- \equiv \tfrac{1}{2}\begin{pmatrix} 0 & I-N \\ -(I-N) & 0 \end{pmatrix}$$

Now η_+ is symmetric with respect to [,] , while η_- is skew symmetric. Thus one need only impose the hypothesis that there are nonnegative constants β , γ such that

$$|||\eta_- X|||^2 \leqslant \gamma\ |||\eta_+ X|||\ \ |||X||| + \beta\ |||X|||^2 \tag{20}$$

for all $X \in D(\eta)$ in order to insure uniqueness and stability of solutions to the Cauchy problem for (18). (See [1], [2] for details). Supposing (20) to be valid with $X = \begin{pmatrix} x_1 \\ x_2 \end{pmatrix}$ we have, letting $N^\dagger = I-N$, for all $X \in D(\eta)$

$$\|N^\dagger x_1\|^2 + \|N^\dagger x_2\|^2 \leqslant \gamma\,(\|N^\dagger x_2\|^2 + \|N^\dagger x_1 + Mx_2\|^2)^{\frac{1}{2}}(\|x_1\|^2 + \|x_2\|^2)^{\frac{1}{2}}$$
$$+ \beta\,(\|x_1\|^2 + \|x_2\|^2)$$

Letting $x_1 = x$, $x_2 = 0$, we obtain, for any $\alpha > 0$,

$$\|N^{\dagger}x\|^2 \leqslant \gamma \|N^{\dagger}x\| \|x\| + \beta\|x\|^2$$

$$\leqslant \frac{\gamma}{2\alpha}\|N^{\dagger}x\|^2 + (\beta + \frac{\gamma\alpha}{2})\|x\|^2$$

Letting $\alpha = (1+\gamma)/2$ so that $0 < \gamma/2\alpha < 1$, we then find that

$$\|N^{\dagger}(t)x\| \leqslant [(1 + \gamma)(\beta + \frac{\gamma\alpha}{2})]^{\frac{1}{2}}\|x\| , \tag{21}$$

which says that $N^{\dagger}$ and therefore $N = I - N^{\dagger}$ is bounded. But this is just the content of (N-I) in case $P \equiv I$. This, we are forced to deal with the equation directly if we hope to allow both M and N to be unbounded operator families. This is the content of the next section.

III. THE GENERAL CASE

In this section we consider the questions of uniqueness and stability for the linear version of Problem III. By stability here, we mean the notion of Hölder stability in the sense of F. John [4], which has found wide application (see [5], [6], [12] and [13] for example). The method is as follows : Let $w = u_1 - u_2$ where u_1, u_2 are solutions of Problem III with (possibly) different initial data.

We construct a realvalued functional $F(w(t))$ with the following properties: (i) $F(w(t_o)) = 0 \Longleftrightarrow w(t_o) = 0$; (ii) $F(w) \geq 0$ and (iii) on every compact subset of $[0,T)$ there are constants k_1 and k_2 such that

$$FF'' - (F')^2 \geq -k_1 FF' - k_2 F^2 \tag{22}$$

(The prime denotes d/dt). It is shown in [1], [2], [9], [10] among other places, that either $F \equiv 0$ or $F > 0$ on $[0,T)$. Moreover, in the latter case, if $[t_o, t_1] \subseteq [0,T)$ and $t_o < t < t_1$ then

$$F(t) \leq L[F(t_o)]^{1-\nu(t)}[F(t_1)]^{\nu(t)} \tag{23}$$

where

$$\ln L = (t_o - t_1)^2 k_1 e^{k_1(t_1 - t_o)} \tag{24}$$

and

$$\nu(t) = \frac{e^{-k_1 t_o} - e^{-k_1 t}}{e^{-k_1 t_o} - e^{-k_1 t_1}} \tag{25}$$

The content of (22) is of course that $\ln F(t) - k_2 t^2$ is a convex function of $e^{-k_1 t}$.

We wish to consider, in this section, the following problem:

Problem III_L

$$\mathcal{L}(t)u \equiv P(t)u_{tt} + M(t)u + N(t)u = \mathcal{F}(t) \qquad 0 \leq t < T ,$$

$$u(0),\ du/dt(0) \ \text{prescribed.}$$

Here $\mathcal{F}(\cdot)$ is a prescribed function (vector valued) which one may suppose simply to be the prescribed value of the operator, $\mathcal{L}(t)u$. Since $\mathcal{L}(t)$ is a linear operator and since the initial conditions are linear, we may suppose that u is the difference of two solutions $u_1, u_2 \varepsilon C^2([0,T);D)$ to $\mathcal{L}(t)u_i = \mathcal{F}_i$ where $\mathcal{F} \equiv \mathcal{F}_1 - \mathcal{F}_2$. We assume that $P(t)$, $M(t)$ and $N(t)$ are defined on a dense domain $D(t) \subseteq H$ and that H is a real Hilbert space. (This last restriction is made simply to avoid making already cumbersome expressions more cumbersome by being forced to write $Re(u,Av)$ in place of (u,Av)). We shall assume that whenever $v,w \varepsilon C^1([0,T);D)$ and $A(t)$ is a symmetric operator for each t, the scalar valued function

$$f(t) \equiv (v(t),A(t)w(t))$$

is continuously differentiable. We shall denote the quantity $f'(t) - (v_t,Aw) - (v,Aw_t)$ by $(v(t),\dot{A}(t)w(t))$. See, for example, [1] and [2] for sufficient conditions on $A(\cdot)$ to insure the existence of $f'(t)$.

We assume that the following hypotheses are satisfied by the the operator families $P(\cdot)$, $M(\cdot)$, $N(\cdot)$:

P-I. For all $t \in [0,T)$ and some $\lambda > 0$, $P(t)$ is symmetric and

$$(x,P(t)x) \geq \lambda(x,x)$$

for all $x \in D(t)$.

P-II. For all $u \in C^1([0,T);D)$[1], $(u(t),P(t)u(t))$ is continuously differentiable and there is a nonnegative function $k(t) \in \mathcal{L}^\infty_{loc}[0,T)$ such that

$$|(u,\dot{P}u)| \leq k(t)(u,Pu).$$

M-I. For each t, $M(t) = M_1(t) + M_2(t)$ where $M_1(t)$ is symmetric and $M_2(t)$ is skew symmetric.

M-II. $M_1(\cdot)$ satisfies P-I for some $\tilde{\lambda} \geq 0$.

M-III. The function (u,M_1u) is continuously differentiable for all $u \in C^1([0,T);D)$ and there are nonnegative $k_1(\cdot),k_2(\cdot) \in \mathcal{L}^\infty_{loc}([0,T))$ such that

$$|(u,\dot{M}_1u)| \leq k_1(t)(u,M_1(t)u) + k_2(t)(u,P(t)u).$$

M-IV. For each $t \in [0,T)$ and each $x \in D(t)$

$$||M_2(t)x||^2 \leq k_1(t)(x,M_1(t)x) + k_2(t)(x,P(t)x)$$

for some $k_1(\cdot),k_2(\cdot) \in \mathcal{L}^\infty_{loc}[0,T)$ which are nonnegative.

---

1) Here $u \in C^1([0,T);D)$ means that $u : [0,T) \longrightarrow H$ is differentiable in the strong sense and that for each $t \in [0,T)$, $u(t), du(t)/dt \in D(t)$.

One of the following sets of hypotheses hold for the operator family $N(\cdot)$:

N-I. For each t, $N(t)$ is symmetric.

N-II. There is a real constant δ (possibly ≤ 0) and nonnegative $\gamma_1(\cdot),\gamma_2(\cdot)\varepsilon\mathcal{L}^{\infty}_{loc}([0,T))$ such that (u,Nu) is continuously differentiable and

$$(u,\dot{N}u) \geq -\delta(u,Nu) - \gamma_1(t)(u,M_1u) - \gamma_2(t)(u,Pu).$$

N'-I. For each t, $N(t) = N_1(t) + N_2(t)$ where $N_2(t)$ is skew symmetric and

N'-II. For all $t\varepsilon[0,T)$,

(α) $(x,N_1(t)x) \geq 0$ for all $x\varepsilon D(t)$

or for all $t\varepsilon[0,T)$,

(β) $(x,N_1(t)x) \leq 0$ for all $x\varepsilon D(t)$

N'-III. $N_1(\cdot)$ satisfies N-II above with $\delta \geq 0$ in case N'-II(α) holds or with $\delta \leq 0$ if N'-II(β) holds.

N'-IV. For each $t\varepsilon[0,T)$ some constant $\gamma \geq 0$ and all $x\varepsilon D(t)$

$$||N_2x||^2 \leq \gamma|(x,N_1x)| + \gamma_1(t)(x,M_1(t)x) + \gamma_2(t)(x,P(t)x)$$

for some nonnegative $\gamma_1(\cdot),\gamma_2(\cdot)\varepsilon\mathcal{L}^{\infty}_{loc}([0,T))$

Theorem I . Under the preceeding hypotheses, on every compact set K contained in $[0,T)$[2], there exist nonnegative computable R^2, k_1 and k_2 such that

$$F(t) \equiv \int_0^t (u,P(\eta)u)d\eta + \int_0^t (t-\eta)(u,M_1(\eta)u)d\eta + R^2$$

(27)

$$+ (T'-t)(u,Pu)_o + \tfrac{1}{2}[(T')^2 - t^2](u,M_1u)_o$$

satisfies an inequality of the form (22). Here the constants k_1 and k_2 depend only upon bounds for the functions given in the imposed conditions on the operators while $T' = \sup\{t \in K\} < T$ and R^2 depends upon the initial data $u(0)$ and $u_t(0)$ as well as the bounds on the operators in the preceeding hypotheses.

Proof : Let $F(t)$ be defined by (27) for any constant R^2 . We shall show how R^2 may be chosen to satisfy the conditions of the theorem. We have

$$F'(t) = (u,P(t)u) - (u,Pu)_o + \int_0^t (u,M_1u)d\eta - t(u,M_1u)_o$$

(28a)

$$= \int_0^t \frac{d}{d\eta}(u,Pu)d\eta + \int_0^t\int_0^\eta \frac{d}{d\tau}(u,M_1u)d\tau d\eta$$

---

[2] The interval $[0,T)$ is chosen without loss of generality. If the problem had been formulated on $[t_o,t_1)$ with initial data at t_o , then, since the translation $t \to t-t_o$ does not affect the operator hypotheses, it is equivalent to the same problem formulated on $[0,t_1-t_o)$ with initial data prescribed at 0 .

$$F'(t) = 2\int_0^t (u,Pu_\eta)d\eta + 2\int_0^t (t-\eta)(u_\eta,M_1u)d\eta$$

(28b)

$$+ \int_0^t (u,\dot{P}u)d\eta + \int_0^t (t-\eta)(u,\dot{M}_1u)d\eta$$

$$\equiv 2I_1 + J_1$$

where J_1 is the sum of the last two integrals in (28b) and $I_1 \equiv \frac{1}{2}(F' - J_1)$. Thus we find that

$$F''(t) = 2\int_0^t (u_\eta,Pu_\eta)d\eta + 2\int_0^t (t-\eta)(u_\eta,M_1u_\eta)d\eta + 2(u,Pu_t)_o +$$

(29)

$$2\int_0^t (u,Pu_{\eta\eta})d\eta + 2\int_0^t (t-\eta)(u,M_1u_{\eta\eta})d\eta + 2t(u_t,M_1u)_o +$$

$$+ \frac{dJ_1}{dt} + 2\int_0^t [(u_\eta,\dot{P}u) + (t-\eta)(u_\eta,\dot{M}_1u)]d\eta \quad .$$

Letting $L(t)$ denote the last integral on the right of (29), we have

(30) $$F^2(\ln F)'' = FF'' - (F')^2 = 4S^2 + 4FQ^2 + 2FE(t) + FJ_1'$$

$$+ FD_1 - 4I_1J_1 - J_1^2 + 2FL(t)$$

where we have set

(31) $$Q^2 \equiv R^2 + (T'-t)(u,Pu)_o + \tfrac{1}{2}((T')^2-t^2)(u,Mu)_o \quad ,$$

(32) $$E(t) \equiv \int_0^t [(u,Pu_{\eta\eta}) - (u_\eta,Pu_\eta)]d\eta$$

$$+ \int_0^t (t-\eta)[(u,M_1u_{\eta\eta}) - (u_\eta,M_1u_\eta)]d\eta$$

(33) $$D_1 \equiv (u,Pu_t)_o + t(u_t,M_1u)_o \quad ,$$

and

(34) $$S^2 \equiv \left(\int_0^t [(u,Pu) + (t-\eta)(u,M_1u)]d\eta \int_0^t [(u_\eta,Pu_\eta) + (t-\eta)(u_\eta,M_1u_\eta)]d\eta \right) - \left(\int_0^t [(u,Pu_\eta) + (t-\eta)(u,M_1u_\eta)]d\eta \right)^2 .$$

The remainder of the proof will consist in showing that all of the terms following $4S^2$ on the right of (30) can be bounded below by an expression of the form $-k_1F^2 - k_2FS - k_3FF'$ where k_1,k_2 and k_3 are computable, nonnegative constants, ("Computable" is defined below). This leads to, upon completion of squares,

(22) $$FF'' - (F')^2 \geq - k_1FF' - k_2F^2$$

where k_1 and k_2 are computable and nonnegative.

We record, at this point, some inequalities which will be useful in the sequel. We shall denote generic, computable, nonnegative constants by subscripted k's and l's . Here, "computable" means that the constants are known functions of the parameters appearing in the hypotheses on the operator families as well as any coefficients used in applications of the arithmetic-geometric mean inequality and (possibly) T' . Constants which depend upon these computable constants and the initial data or the value of the operator ($\mathcal{Y}$) will be denoted by subscripted d's .

We first observe that

$$|F'(t)| \leq (u,P(t)u)_o + t(u,M_1u)_o + (u,P(t)u) + \int_0^t (u,M_1u)d\eta$$

However, from (28a) the sum of the last two terms on the right is simply the sum of the first two and $F'(t)$ so that

$$(35) \qquad |F'(t)| \leq 2d_1 + F'(t)$$

(Here $d_1 = (u,P(t)u)_o + T'(u,M_1u)_o \geq 0$) .

Now we know that, from (34) ,

$$\left(\int_0^t [(u,Pu) + (t-\eta)(u,M_1u)]d\eta \int_0^t [(u_\eta,Pu_\eta) + (t-\eta)(u_\eta,M_1u_\eta)]d\eta\right)^{\frac{1}{2}}$$

$$= (S^2 + I_1^2)^{\frac{1}{2}}$$

$$= [S^2 + \tfrac{1}{4}(F' - J_1)^2]^{\frac{1}{2}}$$

$$\leq S + \tfrac{1}{2}|F'| + \tfrac{1}{2}|J_1| .$$

Now, applying Hypotheses P-I, P-II, M-I, M-II and M-III we see that $|J_1| \leq k_1(F-R^2)$ for some computable, positive, constant k_1. It therefore follows after using (35) that for some nonnegative computable constants k_i with $k_5 > 0$,

$$(36) \quad \max \left\{ \begin{array}{l} \left[\int_0^t (u,Pu)d\eta \int_0^t (u_\eta,Pu_\eta)d\eta\right]^{\frac{1}{2}} \\ \left[\int_0^t (u,Pu)d\eta \int_0^t (t-\eta)(u_\eta,M_1u_\eta)d\eta\right]^{\frac{1}{2}} \\ \left[\int_0^t (u_\eta,Pu_\eta)d\eta \int_0^t (t-\eta)(u,M_1u)d\eta\right]^{\frac{1}{2}} \\ \left[\int_0^t (t-\eta)(u,M_1u)d\eta \int_0^t (t-\eta)(u_\eta,M_1u_\eta)d\eta\right]^{\frac{1}{2}} \end{array} \right\} \leq k_1F+k_2F'+k_3S + k_4d_1-k_5R^2$$

Now we go to work on the right hand side of (30), term by term. Since

$$|J_1| \leq k_1(F-R^2) \leq k_1F$$

we have

$$0 \leq J_1^2 \leq k_1^2(F-R^2)^2 \leq k_1F^2 - k_2FR^2$$

as $0 \leq F - R^2 \leq F$. Now

$$I_1 \equiv \int_0^t (u,Pu_\eta)d\eta + \int_0^t (t-\eta)(u,M_1u_\eta)d\eta \quad .$$

Therefore

$$|I_1| \leq \int_0^t |(u,Pu_\eta)|d\eta + \int_0^t (t-\eta)|(u,M_1u_\eta)|d\eta \quad .$$

Whence

$$(38)\qquad |I_1| \leq \left[\int_0^t (u,Pu)d\eta \int_0^t (u_\eta,Pu_\eta)d\eta\right]^{\frac{1}{2}} + \left[\int_0^t (t-\eta)(u,M_1u)d\eta \int_0^t (t-\eta)(u_\eta,M_1u_\eta)d\eta\right]^{\frac{1}{2}}$$

$$\leq k_1F + k_2F' + k_3S + k_4d_1 - k_5R^2$$

using (36) and (37). Therefore, there are computable nonnegative constants k_i, $i = 1,..,4$ with $k_4 > 0$ such that

$$(39)\qquad J_1^2 + |4I_1J_1| \leq k_1F^2 + k_2FS + k_3FF' + k_4Fd - k_5FR^2$$

Now

$$J_1' = (u,\dot{P}u) + \int_0^t (u,\dot{M}_1u)d\eta$$

so that from P-I, P-II and M-II we have

$$(40)\qquad |J_1'| \leq k[(u,Pu) + \int_0^t (u,M_1u)d\eta + \int_0^t (u,Pu)d\eta]$$

$$\leq k_1F + k_2F' + k_3d_1$$

Also, it is easily seen that $|L(t)| \leq |I_1(t)|$[3].

We therefore have the following estimate if we combine (37), (38), (39) and (40) and use the facts that $|L| \leq |I_1|$ and that

$$d_2 \equiv |(u,Pu_t)_o| + T'|(u_t,M_1u)_o|$$

is an upper bound for D_1 and is only data :

(41)

$$FJ_1' + FD_1 - 4I_1J_1 - J_1^2 + 2FL \geq F(l_3R^2 - l_1d_1 - l_2d_2) - k_1F^2 - k_2FS - k_3FF'$$

where the k's and l's are nonnegative computable constants and $l_3 > 0$.

We now try to find a lower bound on the term $FE(t)$ of the form $-k_1F^2 - k_2FS - k_3FF' + DF$ where D is a data term. We begin by observing that

(42)
$$E(t) \equiv -\int_0^t [(u,M_1u_\eta) + (u,M_2u_\eta) + (u,Nu) + (u_\eta,Pu_\eta)]d\eta$$
$$+\int_0^t (t-\eta)(u,M_1u_{\eta\eta})d\eta - \int_0^t (t-\eta)(u_\eta,M_1u_\eta)d\eta + \int_0^t (u,\mathcal{F})d\eta .$$

However,

$$\int_0^t (u,M_1u_\eta)d\eta = \int_0^t (t-\eta)\frac{d}{d\eta}(u,M_1u_\eta)d\eta + t(u,M_1u_t)_o$$
$$= \int_0^t (t-\eta)(u,M_1u_{\eta\eta})d\eta + \int_0^t (t-\eta)(u_\eta,M_1u_\eta)d\eta$$
$$+\int_0^t (u,\dot{M}_1u_\eta)d\eta + t(u,Mu_t)_o .$$

It follows that

$$(43)\qquad E(t) = -\int_0^t [(u,Nu) + 2(t-\eta)(u_\eta,M_1u_\eta) + (u_\eta,Pu_\eta)]d\eta$$

$$+\int_0^t (u,M_2u_\eta)d\eta + \int_0^t (u,\mathfrak{F})d\eta - \int_0^t (u,M_1u_\eta)d\eta + D_3$$

where D_3 is a data term. We treat the last three integrals on the right of (43) as follows :

$$\int_0^t (u,M_2u_\eta)d\eta \geq -\int_0^t \|u\|\,\|M_2u_\eta\|d\eta$$

$$(44)\qquad \geq -\int_0^t \|u\|\,[k_1(u_\eta,M_1u_\eta)^{\frac{1}{2}} + k_2(u_\eta,Pu_\eta)^{\frac{1}{2}}]d\eta \quad \text{(M-IV)}$$

$$\geq -k_1\int_0^t (u,Pu)^{\frac{1}{2}}(u_\eta,M_1u_\eta)^{\frac{1}{2}}d\eta - k_2\int_0^t (u,Pu)^{\frac{1}{2}}(u_\eta,Pu_\eta)^{\frac{1}{2}}d\eta$$

$$\geq -k_1\left(\int_0^t (u,Pu)d\eta \int_0^t (u_\eta,M_1u_\eta)d\eta\right)^{\frac{1}{2}} -$$

$$-k_2\left(\int_0^t (u,Pu)d\eta \int_0^t (u_\eta,Pu_\eta)d\eta\right)^{\frac{1}{2}}$$

$$\geq -k_1F - k_2S - k_3F' + k_4d_1 - k_5R^2 \quad [(36)]\ .$$

$$(45)\qquad \int_0^t (u,\mathfrak{F})d\eta \geq -\int_0^t \|u\|\,\|\mathfrak{F}\|d\eta$$

$$\geq -k_1\int_0^t (u,Pu)d\eta - k_2\int_0^t (\mathfrak{F},\mathfrak{F})d\eta$$

$$\geq -k_1F - k_2d_4\ \cdot\ (d_4 \equiv \int_0^{T'} (\mathfrak{F},\mathfrak{F})d\eta)\ .$$

$$(6) \qquad \int_0^t (u,\dot{M}_1 u_\eta)d\eta \leq \int_0^t |(u,\dot{M}_1 u_\eta)|d\eta$$

$$\leq k_1 \left(\int_0^t (u,M_1 u)d\eta \int_0^t (u_\eta,M_1 u_\eta)d\eta \right)^{\frac{1}{2}}$$

$$+ k_2 \left(\int_0^t (u,Pu)d\eta \int_0^t (u_\eta,Pu_\eta)d\eta \right)^{\frac{1}{2}} \quad 3)$$

$$k_1 F + k_2 F' + k_3 S + k_5 d_5 - k_4 R^2 \qquad |(36)|.$$

(Here again, all the k's are generic, nonnegative computable constants.) Thus, for some constants k_i with $k_i \geq 0$, we have

$$E(t) \geq - \int_0^t |(u_\eta,Pu_\eta) + 2(t-\eta)(u_\eta,M_1 u_\eta) + (u,Nu)|d\eta$$
$$- k_1 F - k_2 F' - k_3 S + k_4 R^2 - k_5 d_5$$

where

---

3) Here we have used the following fact which may be found in [10] and [15]. If A is a symmetric operator on an inner product space V and if B is a positive definite operator on V $(x,Bx)_V > 0$ unless $x = 0$) such that

$$|(Ax,x)_V| \leq |(x,Bx)_V|$$

for all $x \varepsilon V$ then

$$|(Ax,y)_V|^2 \leq (x,Bx)_V (y,By)_V$$

for all $x,y \varepsilon V$.

$$(48) \quad d_5 = l_1(u,Pu)_o + l_2(u_t,Pu_t)_o + l_3(u,M_1u)_o + l_4(u_t,Mu_t)_o$$

$$+ l_5|(u,Nu)_o| + l_6\int_0^{T'} \|\mathfrak{F}\|^2 d\eta$$

and where the l's are nonnegative, computable constants.

Now we define

$$(49) \qquad G(t) \equiv \int_0^t [(u_\eta,Pu_\eta) + 2(t-\eta)(u_\eta,M,u_\eta) + (u,Nu)]d\eta \quad .$$

Assume for the moment the truth of the following

Lemma : Under the hypotheses of the theorem, there are computable nonnegative constants k_i such that

$$(50) \qquad G(t) \leqslant k_1F + k_2S + k_3F' - k_4R^2 + k_5d_5$$

where d_5 is of the form given in (48) with the l's nonnegative and computable.

Combining (41), (44), (45), (46) and the result (50) of the Lemma we obtain

$$F^2(\ln F)'' \geqslant 4S^2 - k_1F^2 - k_2FS - k_3FF' + k_4(R^2-d_5)$$

for some nonnegative computable k_i and a d_5 of the form (48). We now choose R^2 to be this last d_5 and complete the square in $4S^2 - k_2 FS$ to obtain an inequality of the form (22) which is satisfied by $F(t)$ with R^2 of the form (48).

We now prove (50). We have

$$2\int_0^t (t-\eta)(u_\eta, Pu_{\eta\eta} + Mu_\eta + Nu)d\eta = 2\int_0^t (t-\eta)(u_\eta, \mathfrak{F})d\eta .$$

Which yields, after an integration by parts on the left hand side,

$$G(t) = 2\int_0^t (t-\eta)(u_\eta, \mathfrak{F})d\eta - \int_0^t (t-\eta)(u_\eta, \dot{P}u_\eta)d\eta - \int_0^t (t-\eta)(u, \dot{N}u)d\eta$$

$$- t(u_t, Pu_t)_o - t(u, Nu)_o .$$

For the remainder of the argument we shall assume that $N(\cdot)$ satisfies N-I and N-II . The corresponding argument is easily seen to hold if $N(\cdot)$ satisfies the second set of conditions. Using these conditions as well as Schwarz's inequality and P-I , P-II we find that

(51a) $$G(t) \leq k_1\int_0^t (t-\eta)(u_\eta, Pu_\eta)d\eta + k_2\int_0^t (t-\eta)(u, M_1 u)d\eta$$

$$+ \delta\int_0^t (t-\eta)(u, Nu)d\eta + k_3\int_0^t (t-\eta)(u, Pu)d\eta + k_4 d_5$$

(51b)

$$G(t) \leq k_1\int_0^t (t-\eta)(u_\eta, Pu_\eta)d\eta + \delta\int_0^t (t-\eta)(u, Nu)d\eta + k_2 F + k_3 d_5$$

where again the k's are generic, computable and nonnegative while d_5 is of the form (47). We may take $k_1 > 0$ in this last inequality, if it is not already positive.

We also have

$$\int_0^t (t-\eta)(u_\eta,Pu_\eta)d\eta = \int_0^t (t-\eta)\frac{d}{d\eta}(u_\eta,Pu)d\eta - \int_0^t (t-\eta)(u_\eta,\dot{P}u)d\eta$$

$$- \int_0^t (t-\eta)(u,Pu_{\eta\eta})d\eta$$

$$= \int_0^t (t-\eta)(u,M_1u_\eta)d\eta + \int_0^t (t-\eta)(u,M_2u_\eta)d\eta$$

$$+ \int_0^t (t-\eta)(u,Nu)d\eta - \int_0^t (t-\eta)(u,\mathfrak{F})d\eta$$

$$- \int_0^t (t-\eta)(u_\eta,\dot{P}u)d\eta + \text{a data term of the form (48).}$$

In the second of these equations we integrate by parts in the first integral and then use the various hypotheses on $P(\cdot)$ and $M(\cdot)$ to obtain an inequality of the following form :

$$\int_0^t (t-\eta)(u_\eta,Pu_\eta)d\eta \leq \int_0^t (t-\eta)(u,Nu)d\eta + k_1\left[\int_0^t (u,M_1u)d\eta \int_0^t (u_\eta,Pu_\eta)d\eta\right]^{\frac{1}{2}}$$

$$+ k_2\left[\int_0^t (u,Pu)d\eta \int_0^t (u_\eta,Pu_\eta)d\eta\right]^{\frac{1}{2}} + k_3\int_0^t (u,Pu)d\eta \;.$$

(52)

$$+ k_4\left[\int_0^t (u,M_1u)d\eta \int_0^t (u_\eta,M_1u_\eta)d\eta\right]^{\frac{1}{2}} + \text{a data term of the form (48).}$$

For any $k \geq k_1$ where k_1 is the coefficient of the first integral in (51b), and for any $\alpha \in (\frac{1}{2},1]$ we have

$$(53)\quad G(t) \leq (1-\alpha)k \int_0^t (t-\eta)(u_\eta,Pu_\eta)d\eta + \alpha k \int_0^t (t-\eta)(u_\eta,Pu_\eta)d\eta$$

$$+ \delta \int_0^t (t-\eta)(u,Nu)d\eta + k_2F + k_3d_5 .$$

We now choose k so large that $k \geq k_1$ and $0 < (1-\alpha)k < \alpha k + \delta \equiv r$. Using (52) to estimate the second integral on the right of (53) and the positive definiteness of M_1 we obtain,

$$G(t) \leq r \int_0^t G(\tau)d\tau + I(t)$$

where $I(t)$ is a nonnegative increasing function of t . Thus, after a quadrature,

$$(54)\qquad G(t) \leq e^{rt}I(t) \leq e^{rT'}I(t) .$$

Therefore, taking into consideration the form of $I(t)$. (It is a linear combination of the last five terms on the right of (52) and $F(t)$ with nonnegative, computable coefficients) and (36) we obtain the statement (50) of the Lemma from (54) .

<u>Corollary 1</u> . Let $v_1,v_2 \in C^2([0,T);D)$ be solutions on $[0,T)$ to Problem III_L with $\mathcal{L}(t)v_i = \mathfrak{F}(t)$ and $v_i(0) = u_o$, $v_{i,t}(0) = u_1$. If

the operator coefficients satisfy the hypotheses of the previous theorem then $v_1(t) = v_2(t)$ for all $t \varepsilon [0,T)$.

Proof : Let $u = v_1 - v_2$. Then $\mathcal{L}(t)u = 0$ and $u(0) = u_t(0) = 0$. Thus, $F(t)$ as given by (27) and (48) ($R^2 \equiv d_5$) reduces to

$$F(t) = \int_0^t (u,Pu)d\eta + \int_0^t (t-\eta)(u,M_1u)d\eta$$

and satisfies an inequality of the form (22). Since, in this case, $F(0) = 0$, we have $F(t) \equiv 0$ so that $(u,Pu) \equiv 0$ and $(u,M_1u) \equiv 0$. Therefore, using the positive definiteness of either $P(t)$ or $M(t)$ for each $t \varepsilon [0,T)$, $u(t) = 0$ or $v_1(t) = v_2(t)$.

Corollary 2 . Under the hypotheses of the preceeding theorem, the solutions to Problem III_L are Hölder stable in the sense of F. John [4] on compact subsets of $[0,T)$.

Proof : By (23) we see that for $0 \leqslant t \leqslant T' - \mu < T' < T$ we have, with $t_o = 0$ and $t_1 = T'$

$$\int_0^t (u,Pu)d\eta + \int_0^t (t-\eta)(u,M_1u)d\eta \leqslant L[F(0)]^{1-\delta}[F(T')]^{\delta}$$

where $u = v_1 - v_2$ is the difference of two solutions to the problems $\mathcal{L}(t)v_i = \mathcal{F}_i$ in $[0,T)$ $v_i(0)$, $v_{i,t}(0)$ prescribed. Here $\mathcal{F} = \mathcal{F}_1 - \mathcal{F}_2$ and $0 < \delta \equiv \nu(T'-\mu) < 1$ where ν is given by (25). Now suppose $\mathcal{F}$, $u(0)$ and $u_t(0)$ are small in the sense that

$$\int_0^T \|\mathcal{F}\|^2 d\eta, (u,Pu)_o, (u_t,Pu_t)_o, (u,M_1u)_o, (u_t,M_1u_t)_o, |(u,Nu)_o|$$

are all small. Then as

$$F(0) = T'(u,Pu)_o + \tfrac{1}{2}T'^2(u,M_1u)_o + R^2$$

where R^2 is of the form (48), we see that $F(0)$ is of the form (48) as well. Thus, if the above quantities are all less than some prescribed ε^2, then, $F(0) \leqslant k\varepsilon^2$ for some computable constant k and, for $t \in [0,T'-\mu]$ we have

$$(55) \qquad \int_0^t (u,Pu)d\eta + \int_0^t (t-\eta)(u,M_1u)d\eta \leqslant K\varepsilon^{2-2\delta}[F(T')]^\delta$$

where

$$F(T') = \int_0^{T'} (u,Pu)d\eta + \int_0^{T'} (T'-\eta)(u,M_1u)d\eta + R^2 .$$

Thus if we consider only those solutions which lay in some a priori bounded class, say

$$\mathcal{M} = \{v \in C^2([0,T),D) \,|\, \int_0^T (v,Pv) + (T-\eta)(v,M_1v)d\eta \leqslant M^2\}$$

where M is an a priori given constant. Then for $t \in [0,T' - \mu]$, we have

$$(56) \qquad \int_0^t (u,Pu)d\eta + \int_0^t (t-\eta)(u,M_1u)d\eta \leq KM^{2\delta}\ \varepsilon^{2(1-\delta)} \quad .$$

Thus we have demonstrated Hölder stability in the sense of F. John [4].

The corresponding nonlinear problem

$$(56) \qquad \mathcal{L}(t)u = Pu_{tt} + Mu_t + Nu = f(t,u,u_t) + \mathcal{F}$$

$$u(0), u_t(0) \text{ prescribed}$$

is easily treated provided we assume that for $u^1, u^2 \in C^2([0,T);D)$ we have, for some nonnegative λ_i, μ_i

$$\|f(t,u^1,u_t^1) - f(t,u^2,u_t^2)\|^2 \leq \lambda_1\alpha(t,u) + \lambda_2\alpha(t,u_t) + \lambda_3\beta(t,u)$$

where $u = u^1 - u^2$ and

$$\alpha(t,u) = (u,Pu) + \mu_1 \int_0^t (u,Pu)d\eta$$

$$\beta(t,u) = (u,M_1u) + \mu_2 \int_0^t (u,M_1u)d\eta \ .$$

Let, for $i = 1,2$, u^i be a solution to (56) corresponding to $u^i(0)$, $u_t^i(0)$ and $\mathcal{F}_i$. Let $u = u^1 - u^2$ and note that u solves the problem

(57) $\mathcal{L}(t)u = f^* + \mathcal{F}^*$; $u(0)$, $u_t(0)$ prescribed

where

$$\mathcal{F}^* = \mathcal{F}_1 - \mathcal{F}_2$$

$$f^* = f(t,u^1,u^1_t) - f(t,u^2,u^2_t) .$$

But this can be viewed as just the linear problem with an inhomogeneous term, $\mathcal{F} \equiv f^* + \mathcal{F}^*$ consisting of two pieces, one of which namely $\mathcal{F}^*$, we treated as data in the preceeding theorem. Now inspection of the proof of the last theorem shows that the value of the operator , $\mathcal{F}$, entered into our calculations only as terms of the form $\int_0^t (u,\mathcal{F})d\eta$, $\int_0^t (t-\eta)(u_\eta,\mathcal{F})d\eta$ and $\int_0^t (t-\eta)(u,\mathcal{F})d\eta$ as we see from (43) and the equations preceding (51) and (52) respectively. If in these three integrals we replace $\mathcal{F}$ by f^* , we can, using (36), obtain inequalities of the same form as in (51) and (52). We then complete our proofs as before. The details are omitted.

IV. SOME EXAMPLES

As a first, simple example, we consider the following initial boundary value problem for the equation

$$\mathcal{L}(t)u = \left(-\sum_{i=1}^{n} \frac{\partial}{\partial x_i^2}\right) \frac{\partial^2 u}{\partial t^2}(\vec{x},t) - \left(\sum_{i,j=i}^{i} a_{ij}(\vec{x}) \frac{\partial^2}{\partial x_i \partial x_j}\right) \frac{\partial u}{\partial t}(\vec{x},t) +$$

$$+ \sum_{i,j=1}^{n} \frac{\partial}{\partial x_i} \left[b_{ij}(\vec{x}) \frac{\partial u}{\partial x_j}(\vec{x},t) \right] = 0$$

in a bounded region $\Omega \subseteq R^n$ with boundary Γ smooth enough to admit of applications of the divergence theorem. (Here $u : \bar{\Omega} \times [0,T) \longrightarrow R'$ is taken to be continuous in $\bar{\Omega} \times [0,T)$ and twice continuously differentiable in $\Omega \times [0,T)$. We prescribe

$$u(\vec{x},0)$$
$$\frac{\partial u}{\partial t}(\vec{x},0)$$

for $\vec{x} \in R^n$ while we take as our boundary condition

$$u(\vec{x},t) = 0$$

if $\vec{x} \in \Gamma$ and $0 \leqslant t < T$. This is a natural generalization of the equation studied in [7] and [17]. We shall take the entries of the coefficient matrices (assumed symmetric) $A = [a_{ij}(\vec{x})]$ and $B = [b_{ij}(\vec{x})]$ to be continuously differentiable realvalued functions of the space variable $\vec{x}$ for simplicity. In order to prove uniqueness and stability using the results of the preceeding section, we let

$$\mathcal{H} = \mathcal{L}^2(\Omega)$$

$$\mathcal{D} = \{f \in C^2(\bar{\Omega}) | f = 0 \text{ on } \Gamma\}$$

As our inner product we use the $\mathcal{L}^2(\Omega)$ inner product. For our operators we take

$$Pf = -\Delta_n f \equiv -\sum_{x=1}^{n} \frac{\partial^2 f}{\partial x_i^2}$$

Of course, $\dot{P} = 0$ and

$$(Pf,f) \geqslant \lambda(f,f)$$

for $f \in D$, where $\lambda > 0$ is the first eigenvalue for the (n dimensional) membrane problem. For the operator N, we have

$$(Nf)(\vec{x}) = \sum_{i,j=1}^{n} \frac{\partial}{\partial x_i} \left(b_{ij}(\vec{x}) \frac{\partial f}{\partial x_j}(\vec{x}) \right)$$

for $f \in D$. One easily checks that N is symmetric and that $\dot{N} = 0$.

Let us assume, in addition that the matrix $A = [a_{ij}(\vec{x})]$ has twice continuously differentiable entries. We set, for $f \in D$

$$(Mf)(\vec{x}) = -\sum_{i,j=1}^{n} a_{ij}(\vec{x}) \frac{\partial^2 f}{\partial x_i \partial x_j}(\vec{x}) .$$

Then the operator M decomposes (uniquely) into a symmetric part M_1 whose action is $(f \in D)$

$$(M_1 f)(x) \equiv - \sum_{i,j=1}^{n} \frac{\partial}{\partial x_i}\left(a_{ij}(\vec{x})\frac{\partial f(\vec{x})}{\partial x_j}\right) - \tfrac{1}{2}\left(\sum_{i,j=1}^{n} \frac{\partial^2 a_{ij}(\vec{x})}{\partial x_i \partial x_j}\right)\cdot f(\vec{x})$$

and a skew symmetric part M_2 whose action is

$$(M_2 f)(\vec{x}) \equiv \sum_{i,j=1}^{n} \frac{\partial a_{ij}(\vec{x})}{\partial x_j}\,\frac{\partial f(\vec{x})}{\partial x_i} + \tfrac{1}{2}\left(\sum_{i,j=1}^{n} \frac{\partial^2 a_{ij}(\vec{x})}{\partial x_i \partial x_j}\right)\cdot f(\vec{x})$$

Now if the matrix A is positive definite at each point in D and the smallest eigenvalue has a positive lower bound on Ω and if the coefficient of the undifferentiated term in the definition of M_1 is nonnegative, then it is easily seen that

$$(f, M_1 f) \geqslant \mu(f, Pf) \geqslant \mu\lambda(f,f)$$

for some positive constant μ and all $f \in D$ while $M_1 = 0$. Also, in this case, one easily sees that

$$\|M_2 f\|^2 \leqslant k_1(f, Pf)$$

for a sufficiently large positive constant k_1. Thus, all the conditions of Theorem I are satisfied. But if the matrix A is not positive definite at each point or the coefficient of f in the definition of $M_1 f$ is somewhere negative on Ω, then we can no longer apply our results directly. We can however obtain our uniqueness and stability results in this generality if we can establish the following theorem.

Theorem II . Let $\mathcal{L}(t)u = P(t)u_{tt} + M(t)u_t + N(t)u = \mathcal{F}(t)$ where $\mathcal{F}(t)$ for $t \in [0,T)$ and $u(0)$, $u_t(0)$ are prescribed. Let the family $P(\cdot)$

satisfy P-I and P-II of Theorem I. Suppose that the family $M(\cdot)$ satisfies M-I but that M-II, M-III and M-IV are replaced by

M'-II. There exist constants ρ, $\tilde{\lambda}$ such that

$$(x,(M_1(t)+\rho P(t))x) \geqslant \tilde{\lambda}(x,x)$$

for every $x \in D(t)$ where $\rho > 0$ and $\tilde{\lambda} \geqslant 0$.

M'-III. For all $u \in C^1([0,T);D)$, (u,M_1u) is continuously differentiable and there are $k_1(\cdot),k_2(\cdot) \in \mathcal{L}^\infty_{loc}([0,T))$ and nonnegative such that

$$|(u,\dot{M}_1u)| \leqslant k_1(t)(u,M_1(t)u) + k_2(t)|(u,P(t)u)| .$$

M'-IV. There are nonnegative $k_1(\cdot),k_2(\cdot) \in \mathcal{L}^\infty_{loc}([0,T))$ such that for all $x \in D(t)$ and $t \in [0,t)$

$$\|M_2(t)x\|^2 \leqslant k_1(t)|(x,M_1(t)x)| + k_2(t)(x,P(t)x) .$$

Moreover, assume that for $N(\cdot)$, N-I holds while the inequality in N-II is replaced by

N"-II. $(u,N(t)u) \geqslant -\delta e^{\rho t}(u,N(t)u) - \gamma_1(t)|(u,M_1u)| - \gamma_2(t)(u,Pu)$.

Then this initial value problem has at most one solution.

<u>Proof</u> : The proof of the theorem is very easy. We note that

$$\mathcal{L}(t)u = P(t)(u_{tt} - \rho u_t) + [M(t) + \rho P(t)]u_t + N(t)u = \mathcal{F}(t) .$$

Introducing the change of variables, $\tau = e^{\rho t}$ we have,

$$\mathcal{L}^*(\tau)u^* \equiv \mathcal{L}(t)u = P^*(\tau)u^*_{\tau\tau} + M^*(\tau)u^*_\tau + N^*(\tau)u^* = \mathcal{F}^*(\tau)$$

where

$$P^*(\tau) \equiv e^{2\rho t} P(t)$$

$$M^*(\tau) \equiv e^{\rho t}[M(t) + \rho P(t)]$$

$$N^*(\tau) \equiv N(t)$$

$$\mathcal{F}^*(\tau) \equiv \mathcal{F}(t)$$

and

$$u^*(\tau) \equiv u(t) .$$

The initial value problem becomes, in this notation

$$\mathcal{L}^*(\tau)u^* = \mathcal{F}^* \quad \text{on} \quad [1, e^{\rho T})$$

$$u^*(1),\ u^*_\tau(1) \quad \text{prescribed} .$$

We shall have our uniqueness result if we can show that $P^*(\cdot)$, $M^*(\cdot)$ and $N^*(\cdot)$ satisfy the hypotheses of Theorem I. For P^*, we have for $x \in D(t)$ and $\tau = e^{\rho t}$

$$(x, P^*(\tau)x) = e^{2\rho t}(x, P(t)x) \geq \lambda(x,x) .$$

Using the fact that $d/d\tau = (\rho\tau)^{-1}\, d/dt$ we find

$$(u^*, \dot{P}^*(\tau)u^*) = \rho^{-1}\tau(u, \dot{P}(t)u) + 2\tau^{-1}(u^*, P^*(\tau)u^*)$$

so that

$$|(u^*, \dot{P}^*(\tau)u^*)| \leq \rho^{-1}\tau\, k(t)(u, P(t)u) + 2\tau^{-1}(u^*, P^*(\tau)u^*)$$

$$\leq k^*(\tau)(u^*, P^*(\tau)u^*)$$

where

$$k^*(\tau) \equiv e^{-\rho t}(k(t)/\rho + 2)$$

is locally bounded on $[1, e^{\rho T})$.

For M^* we write $M^* = M_1^* + M_2^*$ where $M_1^*(\tau) = e^{\rho t}(M_1(t) + \rho P(t))$, $M_2^*(t) = e^{\rho t}M_2(t)$ so that M-I holds. Since for $x \varepsilon D(t)$,

$$(x,M_1^*(\tau)x) = e^{\rho t}(x,[M_1(t) + \rho P(t)]x)$$
$$\geqslant \tilde{\lambda}(x,x)$$

M_1^* satisfies M-III . We observe

$$(58) \qquad |(x,M_1(t)x)| \leqslant (x,[M_1(t) + \rho P(t)]x) + \rho(x,P(t)x)$$
$$\leqslant e^{-\rho t}(x,M_1^*(\tau)x) + \rho e^{-2\rho t}(x,P^*(\tau)x) .$$

Thus, for $\tau = e^{\rho t}$ and $x \varepsilon D(t)$,

$$\|M_2^*(\tau)x\|^2 = e^{2\rho t}\|M_2(t)x\|^2$$
$$\leqslant k_1^*(\tau)(x,M_1^*(\tau)x) + k_2^*(\tau)(x,P^*(\tau)x)$$

where

$$k_1^*(\tau) \equiv k_1(t)e^{\rho t}$$
$$k_2^*(\tau) \equiv k_1(t)\rho + k_2(t)$$

are both locally bounded and nonnegative so that M-IV holds. Now,

$$(u^*,\dot{M}_1^*(\tau)u^*) = \tau^{-1}(u^*,M_1^*(\tau)u^*) + \rho^{-1}(u,\dot{M}_1 u) + (u,\dot{P}u) .$$

Whence, from M'-III and (58)

$$|(u^*,M_1^*(\tau)u^*)| \leqslant k_1^*(\tau)(u^*,M_1^*(\tau)u^*) + k_2^*(\tau)(u^*,P^*(\tau)u^*)$$

where

$$k_1^*(\tau) \equiv e^{-\rho t}(1 + k_1(t)/\rho)$$
$$k_2^*(\tau) \equiv e^{-2\rho t}(k_1(t) + k(t) + k_2(t)/\rho)$$

are both locally bounded and nonnegative, so that M-III holds for $M^*(\cdot)$ as well.

Finally, we must check N-I and N-II for $N^*(\cdot)$. Clearly N-I holds. We have, from N"-II and (58),

$$\begin{aligned}(u^*,\dot N^*(\tau)u^*) &= \rho^{-1}\tau^{-1}(u,\dot N(t)u)\\ &\geq -(\delta/\rho)(u^*,N^*(\tau)u^*)\\ &\qquad - \gamma_1(t)|(u,M_1(t)u)| - \gamma_2(t)(u,P(t)u)\\ &\geq -\delta^*(u^*,N^*(\tau)u^*) - \gamma_1^*(\tau)(u^*,M_1^*(\tau)u^*)\\ &\qquad - \gamma_2^*(u^*,P^*(\tau)u^*)\end{aligned}$$

where

$$\delta^* \equiv \delta/\rho \qquad \text{(constant)}$$

$$\gamma_1^*(\tau) \equiv \gamma_1(t)e^{-\rho t}$$

$$\gamma_2^*(\tau) \equiv [\rho\gamma_1(t) + \gamma_2(t)]e^{-2\rho t}$$

so that $N^*(\cdot)$ satisfies N-I and N-II.

<u>Remark</u> : There is an analogous result if the family $N(\cdot)$ satisfies N'-I through N'-IV .

Let us now return to our example. We see that

$$(59) \qquad \int_\Omega f(x)(M_1f)(x)dx = \int_\Omega a_{ij}f_{,i}f_{,j}dx - \tfrac{1}{2}\int_\Omega a_{ij,ij}(x)f^2(x)dx$$

where we have employed the summation convention as well as the notation $f_{,i} \equiv \partial f/\partial x_i$. Let $-\mu$ be a uniform negative lower bound for the smallest eigenvalue of the matrix family $A(x) = (a_{ij}(x))$ in Ω and μ' be a positive upper bound for the coefficient of $f^2(x)$ in the integrand of the second integral in (59). Then

$$(f,M_1f) \geqslant -\mu(f,Pf) - \mu'(f,f) \geqslant -(\mu+\mu'/\lambda)(f,Pf) .$$

Thus with $\rho = \mu + \mu'/\lambda + 1$ and $\tilde{\lambda} = 1$, M'-II is satisfied by M_1 . As for M-III, $\dot{M}_1 = 0$. Since, for some nonnegative computable constants k_i , we have

$$\begin{aligned}\|M_2f\|^2 &= \int_\Omega (a_{ij,j}f,_i + \tfrac{1}{2}a_{ij,ij}f)^2dx \\ &\leqslant k_1\int_\Omega f,_if,_idx + k_2\int_\Omega f^2dx \\ &\leqslant k_3\int_\Omega f,_if,_idx = k_3(f,Pf)\end{aligned}$$

so that M-IV is fulfilled. The family $N(\cdot)$ satisfies N-II as well as N-I if we note that $N = 0$ and let $\delta = \gamma_1 = \gamma_2 = 0$.

In the equation $Pu_{tt} + (M_1+M_2)u_t + Nu = 0$ we had assumed that M_1 and P had the same sign or that M_1 was at least "bounded below" by P when P was nonnegative. Suppose, however, that we have an equation where this is not the case, as for example

$$u_{tt} + u_{xxt} + Nu = 0$$

where N is some differential operator. Here we have that $M = d^2/dx^2$ and $P = I$ are of opposite sign. Then the situation is much more difficult but nevertheless, some statement can be made.

Theorem III . Let $\mathcal{L}(t)u \equiv P(t)u_{tt} + M(t)u_t + N(t)u$. Then under the following hypotheses the problem

$$\mathcal{L}(t)u = 0 \ , \quad u(0) = u_t(0) = 0$$

has only $u \equiv 0$ as a solution in $[0,T)$.

We assume, first of all, that $D(t) = D$ independent of t and that the operator family $P(t)$ is independent of t so that $\dot{P} = 0$. Let the following be fulfilled.

P"-I. P is symmetric and there is a constant $\lambda > 0$ such that for all $x \in D$

$$(Px,x) \geqslant \lambda(x,x) \ .$$

P"-II. The range of P is dense in H .

M"-I. The Cauchy problem

$$\frac{dX(t)}{dt} = P^{-\frac{1}{2}}M(t)P^{-\frac{1}{2}}X(t) \equiv \tilde{M}(t)X(t)$$

$$X(0) = I$$

admits of a solution $X(\cdot) : [0,T) \longrightarrow B(H)$ (the bounded linear operators on H) such that for each t, $X(t)$ is one to one with range contained in D and such that it permutes with $M(t)$. (Here $X'(t)$ is taken in the strong sense.)

M"-II. $\dot{M}(t)$ and $M_1(t)$ exist in the strong sense and $M_1(t)$ satisfies M-III . The differentiation formula

$$d(M(t)v(t))/dt = \dot{M}(t)v(t) + M(t)v_t(t)$$

holds for all $t \in [0,T)$ and $v \in C^1([0,T);D)$.

M"-III. For each $t, M(t) = -M_1(t) + M_2(t)$ where M_2 is skew symmetric and M_1 is symmetric and satisfies, for $t \in [0,T)$ and $x \in D$

$$(x, M_1(t)x) \geqslant \lambda(x,x)$$

for some $\lambda \geqslant 0$.

M"-IV. M_2 satisfies M-IV .

We suppose, in addition that the family $N(\cdot) - M(\cdot)$ satisfies N-I and N-II or N'-I through N'-IV . The following somewhat stringent commutivity hypothesis will also be needed:

C-I. For each $t \in [0,T)$, the bounded operator $X(t)$ permutes with P and $N(t)$. (See Remark III below.)

The proof of this is very easy. Since P is semibounded and symmetric, we can, by the Friedrichs Extension Theorem, take it to be self adjoint. Thus, by the Spectral theorem, $P^{\frac{1}{2}}$ is defined and since $P > \lambda I$ where $\lambda > 0$, P^{-1} and $P^{-\frac{1}{2}}$ both exist and are bounded self adjoint operators. Introducing the change of variables, $v(t) \equiv P^{\frac{1}{2}}u(t)$, the original equation becomes

$$v_{tt} + P^{-\frac{1}{2}}M(t)P^{-\frac{1}{2}}v_t + P^{-\frac{1}{2}}N(t)P^{-\frac{1}{2}}v = 0 .$$

It is not difficult to verify that the operator families

$$\tilde{M}(t) \equiv P^{-\frac{1}{2}}M(t)P^{-\frac{1}{2}} , \quad \tilde{N}(t) = P^{-\frac{1}{2}}N(t)P^{-\frac{1}{2}}$$

satisfy the preceding hypotheses for $P = I$. We thus only need to consider the case $P = I$. Writing M and N for $\tilde{M}$ and $\tilde{N}$ and letting

$$w(t) \equiv X(t)v(t) \tag{60}$$

we obtain successively

$$\begin{aligned} w_t(t) &= X'(t)v(t) + X(t)v_t(t) \\ &= X(t)\{v_t(t) + M(t)v(t)\} \end{aligned}$$

$$\begin{aligned} w_{tt}(t) &= X'(t)\left[v_t(t) + M(t)v(t)\right] \\ &\quad + X(t)\left[v_{tt}(t) + M(t)v_t(t) + M(t)v(t)\right] . \end{aligned}$$

Thus, from M"-I with $P = I$ and C-I, we have

$$w_{tt}(t) - M(t)w_t(t) = X(t)\{-N(t)v(t) + M(t)v(t)\} . \tag{61}$$

Therefore, from C-I and (61) it follows that

$$w_{tt} - M(t)w_t + (N(t) - M(t))w = 0 \tag{62}$$

with initial data $w(0) = w_t(0) = 0$. Now (62) is of the same form as our original equation except that now, the symmetric part of the middle term is $M_1(t)$ and is positive semidefinite by M"-II.

It therefore follows that, since the coefficients of (62) satisfy the hypotheses of Theorem I, we have $w \equiv 0$. Since $X(t)$ is one to one, we obtain $v \equiv 0$ and hence $u \equiv 0$. Q.E.D.

Remark I. Let the Hilbert space H be complex. The existence of an operator family satisfying M"-I is guaranteed by Theorems 3.1

and 3.2 of Friedman [3] provided one assumes in addition that

(i) $P^{-\frac{1}{2}}M(t)P^{-\frac{1}{2}} \equiv \tilde{M}(t)$ is closed

(ii) $(-\infty,0]$ is in the resolvent set of $\tilde{M}(t)$ for each $t \in [0,T)$ and

$$|||R(z,\tilde{M}(t))||| \leq \frac{c}{|z|+1} \quad (z \text{ complex}, \ \operatorname{Re} z \leq 0)$$

(Here $|||\ \ |||$ is the operator norm and $R(z,\tilde{M}) = (zI-\tilde{M})^{-1}$).

(iii) $|||[\tilde{M}(t) - \tilde{M}(\tau)]\tilde{M}^{-1}(s)||| \leq c|t-\tau|^{\alpha} \quad (0 < \alpha < 1)$

(The constants c and α are independent of t, τ, s).

Let M'-III hold with $\lambda > 0$. Now if $\tilde{M}(t)$ is symmetric for each t and if the self adjoint extension (which we know exists from M"-III and Friedrich's extension Theorem [15]) has domain independent of t as well (which we take to be D) then we know that for any z with $\operatorname{Im} z \neq 0$ or $\operatorname{Re} z \leq 0$, $R(z,M(t))$ is defined [15], so that $(-\infty,0]$ is in the resolvent of $\tilde{M}(t)$ for each t. Also, we have for $\tilde{M}$ and $z = -\rho + i\eta$ with $\rho \geq 0$

$$\begin{aligned} |((-zI + M)x,x)| &= |[+\rho(x,x) + (x,Mx)] - i\eta(x,x)| \\ &\geq [(\rho+\lambda)^2\|x\|^4 + \eta^2\|x\|^4]^{\frac{1}{2}} \\ &\geq \tfrac{1}{2}(|z| + \lambda)\|x\|^2 . \end{aligned}$$

Therefore $|||R(z,M(t))||| \leq c(|z|+1)^{-1}$ where $c = \max(2, 2/\lambda)$. Thus, in the case that $\tilde{M}(t)$ is self adjoint, we need only check (iii). In particular, if $\tilde{M}(\cdot)$ is independent of t, (iii) clearly holds. (The existence in this case of $X(t)$ also follows from the

Spectral Theorem. In fact, if $\{E_\mu \mid -\infty<\mu<\infty\}$ is the spectral family associated with M, then

$$X(t) = \int_\lambda^\infty e^{-\mu t} dE_\mu$$

<u>Remark II</u> . In view of Remark I, the most interesting applications of this result occur whenever M is self adjoint, independent of t and strictly negative. In this case $M = 0$ and we may assume either set of hypotheses for $N(\cdot)$ itself. In case, however, that the more general hypotheses of Friedman hold and $\tilde{M}(t)$ is self adjoint with domain independent of t, we can impose N-I and N-II on $N(\cdot)$ and deduce them for $\tilde{N}-\tilde{M}$. As we remarked, if P"-I and P"-II hold we can assume $P = I$. In order to do this, we suppose $\ddot{M}(t)$ exists (again writing M and N for $\tilde{M}$ and $\tilde{N}$) and satisfies M-III . Then $N(t) - M(t)$ is symmetric and, for $x \in D$, $t \in [0,T)$

$$\begin{aligned}(x,[N(t) - \ddot{M}(t)]x) &\geq \delta(x,Nx) - k_1(x,M_1x) - k_2(x,x)\\ &\geq \delta(x,(N-M)x) - |\delta|\ |(x,Mx)|\\ &\qquad - k_1(x,M_1x) - k_2(x,x)\\ &\geq \delta(x,(N-\dot{M})x) - \tilde{k}_1(x,M_1x) - \tilde{k}_2(x,x) .\end{aligned}$$

<u>Remark III</u> . Hypothesis C-I is very strong as it says, roughly, if M and N are differential operators, then the coefficients should not depend on the space variables. However, it may easily be replaced by the somewhat weaker hypothesis.

C'-I. For each $t \in [0,T)$, $X(t)$ permutes with P. Moreover, the operator family $N(\cdot)$ admits of a decomposition of the form

$$N(t) = N^{\dagger}(t) + N^{\dagger\dagger}(t)$$

where $N^{\dagger}(\cdot)$ permutes with $X(t)$ and satisfies the first set of hypotheses on N in Theorem I while $N^{\dagger\dagger}(\cdot)$ satisfies, for all $x \in D$ and $t \in [0,T)$

$$\begin{aligned}\|[N^{\dagger\dagger}(t),X(t)]x\|^2 \leq k_1(t)(y,M_1(t)y) + k_2(t)(y,Py) \\ + k_3(t)\int_0^t (y,M_1(\eta)y)d\eta \\ + k_4(t)\int_0^t (y,Py)d\eta\end{aligned}$$

where the $k_i \in \mathcal{L}^{\infty}_{loc}([0,T))$ and where $y = X(t)x$. ($[A,B]$ denotes the commutator of A and B, namely $AB-BA$.)

This follows immediately from (61), for we then have that

$$w_{tt} - M(t)w_t + (N^{\dagger}(t) - M(t))w = [N^{\dagger\dagger}(t),X(t)]v \equiv f(v)$$

while $\|f(v)\|$ satisfies an estimate of the same form as that considered in the nonlinear version of Theorem I <u>in $w \equiv X(t)v$</u>. Our uniqueness result now follows from this nonlinear version and its corollaries.

We now give some simple examplesto illustrate the previous theory. In some of these examples it might perhaps be easier to employ techniques other than those presented here; however, these examples provide useful illustrations of the Theorem III.

As our first example, let $\Omega = (-\pi,\pi)$ and consider the operator

$$\mathcal{L}(t)u \equiv \frac{\partial^2 u}{\partial t^2} + \frac{\partial^2}{\partial x^2}\left(\frac{\partial u}{\partial t}\right) + a(t)\frac{\partial^2 u}{\partial x^2} + A(t)u$$

where $A(t)$ is a linear operator defined on a domain D_A such that $D_A \subset D_M$ where $D_M = \{f\varepsilon\mathcal{L}^2(-\pi,\pi) \mid f, f'$ are absolutely continuous, $f'' \varepsilon\ {}^2(-\pi,\pi)$ and $f(-\pi) = f(\pi) = 0\}$[4]. (Here $M = -d^2/dx^2$ so that M is self adjoint.) The realvalued function $a(t)$ is assumed to be continuously differentiable and such that $a'(t) \leq -\delta a(t)$ for some constant δ. One can show without much difficulty that

$$X(t)f(x) = \sum_{-\infty}^{\infty} f_n e^{-n^2 t} e^{inx}$$

where the f_n are the Fourier coefficients of $f\varepsilon\mathcal{L}^2(-\pi,\pi)$; so that for $f\varepsilon D_M$

$$X'(t)f = MX(t)f = X(t)Mf \ .$$

For our first choice of $A(t)$ let $a(t) \equiv 0$ and

$$A(t) = A = \sum_{n=0}^{N} \alpha_n(t)\frac{d^{2n}}{dx^{2n}}$$

where the $\alpha_n(t)$ are continuously differentiable and satisfy $(-1)^n \alpha_n'(t) \leq \delta(-1)^{n+1}\alpha_n(t)$. We take as our operator domain

---

4) The equation $u_{tt} + 2\delta u_{xxt} - c^2 u_{xx} = 0$ for constant c, δ is known as Stokes' equation. If $\delta < 0$ then we see that the operator coefficients meet (formally) the conditions of Theorem I while if $\delta \geq 0$, this is a special case of the above equation (after rescaling in x).

$$D = D_A = \{f \in \mathcal{L}^2(-\pi,\pi) \mid f^{(k)}(-\pi) = f^{(k)}(\pi) = 0,\ k = 0,1,2,\ldots,N-1$$

$$\text{and } f^{(2N)} \in \mathcal{L}^2(-\pi,\pi);\ f, f^{(1)},\ldots,f^{(2N-1)}$$

$$\text{absolutely continuous }\}.$$

Then we have uniqueness for the problem

$$\mathcal{L}(t)u = \mathcal{J}(x,t); \quad u_t(x,0),\ u(x,0) \text{ prescribed}$$

$$\left(\frac{\partial}{\partial x}\right)^k u(-\pi,t) = \left(\frac{\partial}{\partial x}\right)^k u(\pi,t) = 0; \quad 0 \leq t < T,\ k = 0,1,\ldots,N-1 .$$

This follows immediately from the last theorem as $X(t)$ and A commute on D_A for all $t \in [0,T)$. (More generally, $-d^2/dx^2 = M$ could be replaced by any polynomial (with constant coefficients) $P(x)$ in M such that $P(n^2) \geq 0$ for $n = 0,1,2,3,\ldots$. Then

$$X(t)f(x) = \sum_{-\infty}^{\infty} f_n\, e^{-P(n^2)t} e^{inx} .$$

Of course, in this case we could also prove uniqueness via use of the (finite) Fourier transform.

As another example, let $a(t)$ satisfy $a'(t) \leq -\delta a(t)$ as above and let p be a fixed integer. Let $\{a_n(t)\}_{-\infty}^{\infty}$ be an infinite sequence of functions on $[0,T)$ such that

$$k(t) \equiv \sup_n |a_n(t)e^{-2pnt}|$$

is in $\mathcal{L}^{\infty}_{loc}([0,T)$. Now define, for each $t \in [0,T)$

$$A_p(t)f(x) \equiv \sum_{-\infty}^{\infty} i(n+p)a_n f_{n+p} e^{inx} .$$

Then, after a routine calculation we find that

$$\|X(t)A_p(t)X^{-1}(t)f - A_p(t)f\| \leqslant -2k(t)(f,Mf)^{\frac{1}{2}} .$$

Thus we see that all the conditions of hypothesis C'-I are satisfied and that there is thus only one solution to the above initial boundary value problem.

Remark : If we consider instead the corresponding initial value problem: $\mathcal{L}(t)u = \mathcal{F}(x,t)$, $u(x,0)$, $u_t(x,0)$ prescribed and in $\mathcal{L}^2(R^1)$, $\mathcal{F}(\cdot,t) \in \mathcal{L}^2(R^1)$, then

$$X(t)f = (2\pi)^{-\frac{1}{2}} \int_{-\infty}^{\infty} e^{-k^2 t}\hat{f}(k)e^{ikx}dx$$

where $\hat{f}$ is the Fourier transform of $f \in \mathcal{L}^2(R^1)$ and where the analog of A_p is

$$A_\rho f(x) = (2\pi)^{-\frac{1}{2}} \int_{-\infty}^{\infty} i(k+\rho)\phi(k,t)\hat{f}(k+\rho)e^{ikx}dk$$

where ρ is any fixed scalar. The (measurable) function $\phi = \phi(k,t)$ is required to satisfy the condition that

$$k(t) \equiv \underset{-\infty<k<\infty}{\text{ess.sup}} |\phi(k,t)|e^{-2\rho kt}$$

is locally bounded on $[0,T)$. We note in addition that if $\phi(\cdot,t) \in \mathcal{L}^2(R^1)$ for each t , then we can put $\hat{\phi}(x,t) = \psi(x,t)$

for each t. Thus, for all f in the domain of d/dx on R^1, we have

$$A_\rho f(x) = (2\pi)^{-\frac{1}{2}} \int_{-\infty}^{\infty} e^{i\rho y} f'(y)\, \psi(x-y,t)dy$$

$$= \left[e^{i\rho y} f' * \psi(\cdot,t)\right](x)$$

where $*$ denotes convolution and $\psi(\cdot,t) \in \mathcal{L}^2(R^1)$. Also, A_ρ does not permute with $X(t)$ except when $\rho = 0$ and is unbounded.(†)

Before closing this section, we should like to make one additional comment. Namely, if $N(t) = b(x)\frac{\partial}{\partial x}$ then, the result of our theorem cannot be used to demonstrate uniqueness. The reason for this is as follows: Define

$$(Bf)(x) = b(x)f(x) .$$

Of course B is a bounded linear operator (assuming $a(x)$ is an essentially bounded function). Computing (formally) $X(t)BX^{-1}(t)f$, we have

$$X(t)BX^{-1}(t)f(x) = \sum_{n=-\infty}^{\infty} \sum_{\ell=-\infty}^{\infty} b_{n-\ell} f_\ell\, e^{(\ell^2-n^2)t}\, e^{inx}$$

However, unless, $b_n = 0$ for all $n \neq 0$,

$$\sum_{n=-\infty}^{\infty} \Big| \sum_{\ell=-\infty}^{\infty} b_{n-\ell}\, f_\ell\, e^{(\ell^2-n^2)t}\Big|^2$$

will not in general be convergent.

As our final example, we consider a function $k(x,y)$ defined almost everywhere on $[-\pi,\pi] \times [-\pi,\pi]$ and square integrable. Let

(†) The author would like to thank Dr. Michael Boon of the Battelle Institute, Advanced Studies Center, for suggesting a version of this example.

$$a_{mn} \equiv (2\pi)^{-2} \int_{-\pi}^{\pi} \int_{-\pi}^{\pi} K(x,y) e^{-imx+iny} dydx$$

denote the Fourier coefficients of K. Define

$$A(t)f(x) \equiv \int_{-\pi}^{\pi} K(x,y)f'(y)dy = i \sum_{m=-\infty}^{\infty} \left(\sum_{n=-\infty}^{\infty} na_{mn} f_n \right) e^{imx}$$

If it happens that

$$k(t) = \sum_n \sum_m |a_{mn}|^2 e^{(n^2-m^2)t}$$

is locally bounded on $[0,T)$, then one can easily show that there is a nonnegative locally bounded function $\ell(t)$ such that

$$\| X(t)AX^{-1}(t)f - Af \|^2 \leq -\ell^2(t)(f,Mf)$$

for all f in the domain of $d^2/dx^2 = M$. We should remark that although a formal integration by parts would indicate that A is a bounded operator, this may not always be possible and A will not, in general, be bounded as we see from the following example : Let $0 < \varepsilon < 1$ be fixed and let

$$a_{mn} = \begin{cases} 0 & |n| > |m| \\ \alpha_{mn} & |n| \leq |m| \end{cases}$$

where the α_{mn} have the property that for each n,

$$\sum_{|m|>|n|} |\alpha_{mn}|^2 = (1+|n|)^{-(1+\varepsilon)}$$

Then, to within an appropriate constant

$$\int_{-\pi}^{\pi}\int_{-\pi}^{\pi} |K(x,y)|^2 dxdy = \sum_{m=-\infty}^{\infty} \sum_{n=-|m|}^{|m|} |\alpha_{mn}|^2 = \sum_{n=-\infty}^{\infty} \frac{1}{(1+|n|)^{1+\varepsilon}} < \infty \quad ,$$

so that the kernel is summable. In addition

$$k(t) = \sum_{m=-\infty}^{\infty} \sum_{n=-|m|}^{|m|} |\alpha_{mn}|^2 e^{2(n^2-m^2)t} \leqslant \int_{-\pi}^{\pi}\int_{-\pi}^{\pi} |K(x,y)|^2 dxdy \quad .$$

However, if A were a bounded operator, then as is well known,

$$\alpha \equiv \sup_n \left(\sum_{m=-\infty}^{\infty} n^2 |a_{mn}|^2 \right) < \infty \ .$$

(For an infinite array (b_{mn}) to represent a bounded linear transformation in ℓ^2, it is <u>necessary</u> that $\sup_n \sum_m |b_{mn}|^2 < \infty$). But for large $|n|$,

$$|n|^2 \sum_{|m|>|n|} |a_{mn}|^2 \approx |n|^{1-\varepsilon} \longrightarrow \infty$$

as $|n| \longrightarrow \infty$. Moreover, this operator A, does not commute with $X(t)$ for any $t \in (0,T)$.

V. SOME RELATED THEOREMS

In this section we wish to mention some related works of Agmon [1] and Lions [11] and to compare our theorems with those of these authors.

Agmon [1], in the course of his work on questions of uniqueness of solutions to the Cauchy problem for elliptic equations with nonanalytic coefficients and questions of unique continuation of solutions to such equations, proved the following abstract theorem whose principle application is the Aronszajn-Cordes theorem (this theorem says that a solution to a second order elliptic equation $\mathcal{L}u = 0$ on some open connected subset $\Omega \subseteq R^n$ in which the coefficients of the principle part are $C^1(\Omega)$ and the other coefficients are in $\mathcal{L}^{\infty}_{loc}(\Omega)$ and which has a zero of infinite order in Ω must be identically zero in Ω) :

Theorem : (Agmon [1]. We state the theorem using the notation of Agmon). Let $u : [0,T) \longrightarrow H$ be a solution to the following differential inequality :

$$\left\|\frac{d^2u}{dt^2} + 2L(t)\frac{du}{dt} + B(t)u\right\| \leqslant \gamma_1(t)\left\{\left\|\frac{du}{dt}\right\|^2 + Q_t(u,u)\right\}^{\frac{1}{2}} + (\gamma_0(t))^2\|u\|$$

in an interval $[0,T)$ where $Q_t(x,x)$ is a Hermitian, positive semidefinite quadratic form. Then, under the hypotheses listed below, if $u(\bar{t}) = du/dt(\bar{t}) = 0$ for some $\bar{t} \in [0,T)$, then $u \equiv 0$ in $[0,T)$.

i. $D_{B(t)} = D_2 \subseteq D_{L(t)} = D_1$ and both are independent of $t \in [0,T)$.

ii. γ_o and γ_1 are continuous, nonincreasing functions of $t \in [0,T)$ and are in $\mathcal{L}^1([0,T))$.

iii. Q_t is defined on D_1 for each $t \in [0,T)$ and $Q_t(x,x) \geq 0$ for all $x \in D_1$.

iv. There is a $\Theta \in (0,1)$ such that for all $x \in D_1$ and $t \in [0,T)$

$$\|L(t)x\|^2 \leq \Theta^2 Q_t(x,x) + (\gamma_o(t))^2\|x\|^2 .$$

v. For all $x \in D_2$, $y \in D_1$ and $t \in [0,T)$

$$|(B(t)x,y) + Q_t(x,y)| \leq \gamma_1(t)(Q_t(x,x))^{\frac{1}{2}}\|y\| + (\gamma_o(t))^2\|x\|\,\|y\|$$

vi. For each $x \in D_1$, $Q_t(x,x)$ is in $C^1[0,T)$ and

$$\left|\frac{d}{dt} Q_t(x,x)\right| \leq \gamma_1(t)Q_t(x,x) + (\gamma_o(t))^2\|x\|^2$$

vii. For each $x,y \in D_1$ and for each $t \in [0,T)$,

$$\|(Lx,y) + (x,Ly)\| \leq \gamma_1(t)\|x\|\,\|y\| .$$

viii. For each $x \in D_1$, $L(t)x \in C^1([0,T),H)$ and,

$$\|\dot{L}(t)x\| \leq \gamma_1(t)\sqrt{Q_t(x,x)} + (\gamma_o(t))^2\|x\| .$$

ix. For every $x \in D_2$

$$|\mathrm{Re}(B(t)x, L(t)x)| \leq \gamma(t) Q_t(x,x) + (\gamma_o(t))^2 \|x\|^2 .$$

x. For every solution u of the differential inequality $Q_t(u,u) \in C^1([0,T))$ and $L(t)u(t) \in C^1([0,T);H)$ and the following differentiation formulae hold :

$$\frac{d}{dt}(Q_t(u,u)) = \dot{Q}_t(u,u) + Q_t(\dot{u},u) + Q(u,\dot{u})$$

$$\frac{d}{dt} L(t)u(t) = \dot{L}(t)u(t) + L(t)\dot{u}(t) \quad .$$

(Agmon also shows that if $T = +\infty$ and $\|u\|$ is bounded then, for some $t_o > 0$, $u(t_o) = 0$ implies $u(t) \equiv 0$ for $t > t_o$, and that $\|u\|^2$ satisfies, for $t \geq t_o$, a convexity inequality).

In general, Theorem 1 and the above theorem of Agmon are not equivalent nor does either imply the other. Specifically, while (vii) above implies that the symmetric part of L be bounded, we required it (M_1 in our notation) to be positive definite. Moreover, while $N(t)$ was not required to be bounded in terms of a positive semidefinite quadratic form Q_t as is $B(t)$ in (v) and (ix) above we could not include a term of the form $|(u,Nu)|$ in the assumed estimate on $\mathcal{L}(t)u$. Moreover, the verification of hypotheses (iv) and (ix), as one sees from [1] are somewhat involved in practice. Nevertheless in the important application, they are of course needed and verifiable. These hypotheses allow Agmon to deduce a convexity type inequality for $F(t) = \|u(t)\|^2$ and thus get uniqueness <u>simultaneously</u> for the forward and backward Cauchy problem. Because we assume other hypotheses, which are more readily applicable to partial differential equations of mixed or even no type and are less restrictive we are forced to use the measure

$$F(t) = \int_0^t (u,Pu) + (t-\eta)(u,M_1u)d\eta$$

and thus treat the forward and backward Cauchy problems separately.

Finally, we remark that Lions [11], has proved two uniqueness results, one for the forward Cauchy problem for the equation

$$\frac{d}{dt}\left(P(t)\frac{du}{dt}\right) + M(t)\frac{du}{dt} + N(t)u = 0$$

(Theorem 2.1, p. 156 and Remark 1.1, p. 154 of [11]) and one for the forward Cauchy problem for

$$\frac{d^2}{dt^2}(P(t)u) + \frac{d}{dt}(M(t)u) + N(t)u = 0$$

(Theorem 3.1, p. 160 of [11]). (Lions states and proves his results in the framework of statements about bilinear forms). Without going into details in the interest of brevity, we content ourselves with a few qualitative remarks concerning these two theorems. First of all, built into their hypotheses are sufficient conditions for existence as well. Since the proofs of existence are based on a varient of the lemma of Lax and Milgram ([11] p. 13 and p. 36), the hypotheses contain coercivity conditions on $(x,P(t)x)$, $(x,M(t)x)$ and $(x,N(t)x)$ as well as their t derivatives. These coercivity conditions are then used together with an energy arguement to prove uniqueness. However, these coercivity conditions are in some cases stronger than those which were required in Theorem I (for example in the case of symmetric N , no lower bound for (x,N(t)x) was needed) and in some cases they are simply different.

BIBLIOGRAPHY

[1] S. Agmon, Unicité et convexité dans les problèmes différentiels, Sem. Math. Sup. (1965), Univ. of Montreal Press (1966).

[2] S. Agmon and L. Nirenberg, Lower bounds and uniqueness theorems for solutions of differential equations in a Hilbert space, Comm. Pure Appl. Math. 20 (1967) 207-229.

[3] A. Friedman, Partial Differential Equations, Holt, Rinehart and Winston, Inc. (1969).

[4] F. John, Continuous dependence on data for solutions of partial differential equations with a prescribed bound, Comm. Pure Appl. Math. 13 (1960) 551-585.

[5] R.J. Knops and L.E. Payne, Stability in linear elasticity, Int. Journal of Solid Structures 4 (1968) 1233-1242.

[6] R.J. Knops and L.E. Payne, On uniqueness and continuous dependence in dynamical problems of linear thermoelasticity, Int. Journal of Solid Structures 6 (1970) 1173-1184.

[7] Y.Y. Kim, On the spectrum of an operator connected with a mixed problem, Differentrial'nye Uravneniya 3 (1967) 109-113.

[8] H.A. Levine, On a theorem of Knops and Payne in dynamical linear thermoelasiticity, Arch. Rat. Mech. Anal. 38 (1970) 290-307.

[9] H.A. Levine, Logarithmic convexity, first order differential inequalities and some applications, Trans. Am. Math. Soc. 152 (1970) 299-320.

[10] H.A. Levine, Logarithmic convexity and the Cauchy problem for some abstract second order differential inequalities, J. Differential Equations 8 (1970), 34-55.

[11] J.L. Lions, Equations Différentiel es et Opérationelles et Problèmes aux Limites, Springer Verlag (1961).

[12] L.E. Payne, Bounds in the Cauchy problem for the Laplace equation, Arch. Rat. Mech. Anal. 5 (1960) 35-45.

[13] L.E. Payne, On some nonwell posed problems for partial differential equations, Proc. Adv. Sympos. Numerical Solutions of Nonlinear Differential Equations (Madison Wisc. 1966) Wiley, N.Y. (1966) 239-263.

[14] L.E. Payne and D. Sather, On some improperly posed problems for quasilinear equations of mixed type, Trans. Amer. Math. Soc. 128 (1967) 135-141.

[15] F. Riesz and B. Sz.-Nagy, Functional Analysis, Ungar Pub. Co. (1961).

[16] S.L. Sobolev, 48 Reunione Società Italiana per il progresso della Scienza, Roma (1965) 192-208.

[17] T.S. Zelenjak, A mixed problem for an equation which cannot be solved for the highest order time deviative, DAN, SSSR (6) (1965) 1225-1228.

STABILIZED QUASI-REVERSIBILITE AND OTHER NEARLY-BEST-POSSIBLE METHODS FOR NON-WELL-POSED PROBLEMS*

K. MILLER

We have been seeing throughout this symposium that restriction of attention to the class of solutions satisfying a prescribed global bound stabilizes many of the ill-posed problems of partial differential equations. This concept, incidentally, was the contribution of Carlo Pucci [11], [12]. Fritz John's influential paper [4] enlarged upon this concept and helped generate much of the new interest in these problems.

The study of ill-posed problems with a prescribed bound seems to divide quite naturally into two separate tasks:-

(i) Derivation of stability estimates.

(ii) Derivation of good numerical methods which somehow incorporate the prescribed bound.

Fortunately there are several quite general numerical methods for linear problems which are "nearly-best-possible" in a certain sense. I mention the following:-

Linear programming methods (L_∞ norm), by Douglas [3] in 1960, and others. (The problem here is usually excessive.)

Methods of partial eigenfunction expansion (L_2 norm), Miller [6] in 1964.

Least squares methods, Miller [7] in 1970. These methods are extremely general and flexible, and numerical if need be; they even allow computer

---

Talk given at the Symposium on Logarithmic Convexity and Non-Well Posed Problems, 21-24 March 1972, Heriot-Watt University, Edinburgh.

*Supported by a C.N.R. Visiting Professorship at Universita di Firenze and by N.S.F. grant

computation of best possible stability estimates. Essentially equivalent methods, but in a different setting and interpretation, were published simultaneously by Backus [2] in 1970, a geophysicist. I also mention a review article [10], to be published in a physical journal, which will explain these L^2 computational methods in a general setting, with nuclear scattering problems as the specific example.

Today, however, I would like to concentrate on some more recent work on stabilized quasi-reversibility (S.Q.R.) methods [8]; these are stabilized versions of the quasi-reversibility methods introduced in the book of Lattes and Lions [5]. I am convinced that methods of this type are extremely limited in applicability, but in certain cases they seem to have very decided computational advantages, allowing us to work problems of a size heretofore impossible.

1. PRELIMINARIES

Consider the ODE on Hilbert space

$$u'(t) = -Au(t),\ t > 0\ , \tag{1}$$

where A is a constant, self-adjoint, non-negative operator. A will usually be unbounded; the important consideration is that its spectral radius $\rho(A)$ is very large.

One typical example is the heat equation on a cylinder,

$$u_t = \Delta u \text{ in } \Omega\times[0,\infty)\ ,$$

$$u = 0 \text{ on } \partial\Omega\times[0,\infty)\ ,$$

where Ω is a bounded domain in R^n. A second example is a difference in χ - differential in t discretization of the first example. We have

$$u'(t) = \Delta_h u(t) \text{ in } \Omega_h\times[0,\infty)\ ,$$

$$u = 0 \quad \text{on } \partial\Omega_h\times[0,\infty)\ ,$$

where u(t) for each t is a discrete function defined only on the points of a grid region Ω_h approximating Ω, $\partial\Omega_h$ is the grid boundary of Ω_h, and Δ_h is the standard five-point finite difference approximation to the Laplacian.

In this case the spectral radius of Δ_h is $O(h^{-2})$. Of course, in these examples we could substitute for Δ and Δ_h any nonpositive self-adjoint elliptic operator L and self-adjoint difference (or finite element) approximation L_h, with coefficients variable with respect to χ but constant with respect to t.

The usual problem for (1) is that of "forward solution" with given initial data $u(0) = \xi \epsilon H$. Then,

$$u(t) = e^{-tA}\xi$$

where e^{-tA} is the semigroup generated by -A (and easily defined in terms of the spectral representation for A).

The "backward problem" is more difficult; given $u(1) = \chi$, the solution (if it exists) must be

$$u(t) = e^{(1-t)A}\chi ,$$

where the unbounded operator $e^{(1-t)A}$ is also easily defined in terms of the spectral representation. However, this is an ill-posed problem; arbitrarily small changes in χ can cause arbitrarily large changes in u(t), and $e^{(1-t)A}\chi$ is not even defined for all χ in H.

Lattes and Lions introduce their method of quasi-reversibility (Q.R.) for this. One first changes the operator A to $A - \delta A^2$, where δ is a small parameter to be chosen appropriately, and solves backwards the problem:-

$$v' = -(A-\delta A^2)v , \quad t \leq 1$$

$$v(1) = \chi .$$

One then lets w(t) be the solution of the unperturbed equation forward from initial values $w(0) \equiv \xi \equiv v(0)$:-

$$w' = -Aw ,$$

$$w(0) = \xi \equiv v(0) .$$

This method has several defects, it seems to me. In the first place, Lattes and Lions ask only that we find a ξ such that $||w(1)-\chi|| < \epsilon$; they do not seek to find an "unknown" solution u(t) for $t < 1$ and in fact the method diverges for $t < 1$. Secondly, some additions to their hypotheses are necessary to even insure $||w(1)-\chi|| < \varepsilon$. Thirdly, the operator $e^{A-\delta A^2}$ which carries χ into ξ has such gigantic norm (like $e^{1/4\varepsilon}$, instead of E/ε which should be possible) that the method is so unstable as to be computationally unfeasible (even for only moderately small values of ε).

Nevertheless, the central idea of the method of Q.R., that of solving (1) backward after first replacing A by a function f(A) with spectrum bounded above, presents several intriguing possibilities, especially from a computational point of view. In general, our construction for a particular choice of f will be as follows: One solves backwards

$$\begin{aligned} v' &= -f(A)v, \quad t \leqslant 1, \\ v(1) &= \chi. \end{aligned} \tag{2}$$

Then, if desired, we solve forward

$$\begin{aligned} w' &= -Aw, \quad t \geqslant 0, \\ w(0) &= \xi \equiv v(0). \end{aligned} \tag{3}$$

2. THE ORIGINAL METHOD OF Q.R.

<u>Problem: Given $\varepsilon > 0$ and χ with $||\chi|| \leqslant 1$, find ξ such that $w(1)$ satisfies</u>

$$||w(1)-\chi|| \leqslant \varepsilon. \tag{4}$$

Well,

$$w(t) = e^{-tA}\xi = e^{-tA}\, e^{A-\delta A^2}\chi = \left(e^{(1-t)A-\delta^2A^2}\chi\right), \quad t \geqslant 0, \tag{5}$$

therefore

$$w(1) - \chi = \left(e^{-\delta A^2} - I\right) .$$

Now Lattes and Lions seem to affirm (incorrectly) that

$$||e^{-\delta A^2} - I|| \leqslant 1 - e^{-\delta\lambda_o^2}, \quad \lambda_o = \inf\{\text{spectrum } A\},$$

and therefore hypothesize that $\lambda_o > 0$ and choose such that $1 - e^{-\delta\lambda_o^2} \leqslant \varepsilon$. For small ε this is essentially equivalent to choosing

$$\delta\lambda_o^2 \approx \varepsilon .$$

This difficulty can be corrected by a modest "smoothness" assumption on χ, for example

$$||(I+A^2)\chi|| \lesssim 1 .$$

Then with $\delta = \varepsilon$ one gets (4) as desired. However, for every instant $t < 1$ the method diverges as $\delta \to 0$. (Here we note that the norm of any function g of A is the maximum of $|g(\lambda)|$ as λ ranges over the spectrum of A; since we know nothing about the spectrum of A except that it is in $[0,\infty)$, we must always consider $\max_{\lambda\varepsilon[0,\infty)} |g(\lambda)|$.) In (5) we have

$$\max_{[0,\infty)} |e^{(1-t)\lambda-\delta\lambda^2}| = e^{(1-t)/4\delta}.$$

In particular, at $t = 0$, this equals $e^{1/4\delta} = e^{1/4\varepsilon}$, which is gigantic for only moderately small ε.

3. S.Q.R. METHODS

We first rephrase the problem, introducing an unknown solution u, and a prescribed bound on u(0). We also recall the basic logarithmic convexity result [1] which stabilizes backward solution of (1) in the presence of a prescribed bound: if y(t) is any solution of (1) (for example the difference between two solutions which both nearly fit the same data χ at $t = 1$ and which both satisfy a prescribed bound at $t = 0$) then $\log||y(t)||$ is convex; hence

$$||y(1)|| \lesssim \varepsilon \text{ and } ||y(0)|| \lesssim E$$

imply that

$$||y(t)|| \lesssim \varepsilon^t E^{1-t} , \text{ for } 0 \lesssim t \lesssim 1 .$$

<u>Problem: Suppose u(t) is an unknown solution of (1) satisfying</u>

$$\left.\begin{array}{ll} ||u(1)-\chi|| \lesssim \varepsilon , & \text{(a)} \\ ||u(0)|| \lesssim E , & \text{(b)} \end{array}\right. \qquad (6)$$

where χ is a given data vector in A, $\varepsilon > 0$ is a known small number, and E is a fixed known number (say on the order of unity). Choose the function F such that our approximation v(t) satisfies

$$||u(t)-v(t)|| \leq 2\,\varepsilon^t E^{1-t} \quad , \; 0 \leq t \leq 1 \; . \tag{7}$$

An error bound of this form is quite a reasonable request. In the first place, it is well known that the stability bound $\varepsilon^t E^{1-t}$ is essentially precise, thus (7) is essentially the best error bound (but for perhaps a factor of two) that could be hoped for. In the second place, it can be recognized that the method of partial eigenfunction expansion can be written in the form (2) with a certain extreme choice of f (see a later section). Since that method itself has the error bound (7), we are assured in advance that (7) is possible for at least one choice of the function f.

We begin derivation of the conditions that f must satisfy by writing

$$\chi = e^{-A}u_o + \psi \; ,$$

where $||u_o|| \leq E$ and $||\psi|| \leq \varepsilon$. Thus

$$v(t) = e^{(1-t)f(A)}\chi =$$

$$= \left[e^{(1-t)f(A)-A}\right]u_o + \left[e^{(1-t)f(A)}\right]\psi$$

therefore

$$||u(t)-v(t)|| \leq \left[e^{-tA}-e^{(1-t)f(A)-A}\right]u_o + \left[e^{(1-t)f(A)}\right]\psi$$

or

$$||B(t)u_o + C(t)\psi|| \leq ||B(t)||E + ||C(t)||\varepsilon \; .$$

Now, in order that each term is less than or equal to $\varepsilon^t E^{1-t}$, it is necessary and sufficient that

$$||B(t)|| \leq (\varepsilon/E)^t \equiv e^{-\alpha t} \qquad \text{(a)}$$

$$||C(t)|| \leq (E/\varepsilon)^{1-t} \equiv e^{\alpha(1-t)} \; , \; 0 \leq t \leq 1 \; , \qquad \text{(b)} \tag{8}$$

where $\alpha = \log(E/\varepsilon)$. Since A can have spectrum anywhere in $[0,\infty)$, we must have

$$\left| e^{-t\lambda} - e^{(1-t)f(\lambda)-\lambda} \right| \lesssim e^{-\alpha t} , \qquad (a)$$

$$\left| e^{(1-t)f(\lambda)} \right| \lesssim e^{\alpha(1-t)} , \text{ for } 0 \lesssim \lambda < \infty , \ 0 \lesssim t \lesssim 1 . \qquad (b) \qquad (9)$$

I will spare you the calculations; we arrive at the result:

Theorem 1. The desired bounds (8a) and (8b) (which imply (7)) hold, for all possible non-negative self-adjoint A, if and only if f satisfies

$$g_1(\lambda) \lesssim f(\lambda) \lesssim G_1(\lambda) , \text{ on } [0,\infty) \qquad (10)$$

where

$$\begin{aligned} G_1(\lambda) &\equiv \min\{\alpha, \lambda + H_1(\lambda)\} , && \lambda\varepsilon[0,\alpha) , \qquad (11) \\ &\equiv \alpha + \log 2 , && \lambda\varepsilon[\alpha,\infty) , \\ g_1(\lambda) &\equiv \lambda - h_1(\lambda) , && \lambda\varepsilon[0,\alpha) , \\ &\equiv -\infty , && \lambda\varepsilon[\alpha,\infty) , \end{aligned}$$

$$H_1(\lambda) \equiv \min_{0\lesssim t\lesssim 1} \left\{ (1-t)^{-1} \log(1+e^{t(\lambda-\alpha)}) \right\} , \ \lambda\varepsilon[0,\alpha) ,$$

$$h_1(\lambda) \equiv \min_{0\lesssim t\lesssim 1} \left\{ -(1-t)^{-1} \log(1-e^{t(\lambda-\alpha)}) \right\} , \ \lambda\varepsilon[0,\infty) .$$

Introducing lower estimates for H_1 and h_1 by minimizing the numerator and maximizing the denominator over t in $[0,1]$, we have the sufficient condition

$$g_2(\lambda) \lesssim f(\lambda) \lesssim G_2(\lambda) , \qquad (12)$$

where G_2, g_2 are defined like G_1, g_2; except with H_1, h_1 replaced by the lower bounds

$$\begin{aligned} H_2(\lambda) &= \log(1+e^{\lambda-\alpha}) \\ h_2(\lambda) &= -\log(1-e^{\lambda-\alpha}) . \end{aligned} \qquad (13)$$

We illustrate this condition (12) in Figure 1.

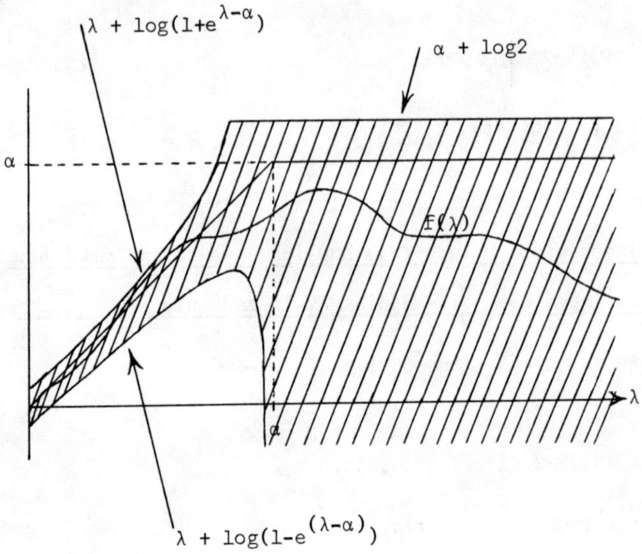

Figure 1. The sufficient condition (12) on f

4. THE "CONTROL PROBLEM" TYPE CRITERION FOR S.Q.R.

We now consider our S.Q.R. methods with respect to the criterion of success (4) used by Lattes and Lions; that is, given χ we wish to find an initial vector or "control" ξ whose exact forward solution w(t) (perhaps this will be physically realized) is close to χ.

Problem: Given χ in H and $\varepsilon > 0$, choose the function f such that $w(t) = e^{-tA}\xi$ satisfies

$$||w(1)-\chi|| \lesssim 2\varepsilon \; . \qquad (a) \qquad (14)$$

There exist of course many ξ satisfying (13). For the sake of optimal stability, we want one with nearly smallest norm. For this reason we hypothesize once again that there exists a solution u of (1) satisfying the bounds (6). We then add the further requirement that our choice of f in

S.Q.R. gives us a w with the "global bound":-

$$||w(0)|| \leq 2E \,. \qquad \text{(b)} \qquad (14)$$

We once again write χ as the sum of its two parts, $e^{-A}u_o$ and ψ. Thus

$$||w(1)-\chi|| = ||I-e^{f(A)-A}\chi||$$

$$= ||\left(I-e^{f(A)-A}\right)e^{-A}u_o + \left(I-e^{f(A)-A}\right)\psi||$$

$$\equiv ||B_1u_o+C_1\psi|| \leq ||B_1||E+||C_1||\varepsilon \,.$$

A sufficient condition for (14) is thus that B_1 and C_1 have norms bounded by $(\varepsilon/E) = e^{-\alpha}$ and 1 respectively. This leads to the following result.

Theorem 2. The desired bounds on $||B_1||$ and $||C_1||$ (which imply (14)) hold, for all non-negative self-adjoint A, if and only if

$$\begin{aligned} &g(\lambda) \leq f(\lambda) \leq G(\lambda) \,, \text{ on } [0,\infty) \,, \\ &G(\lambda) \equiv \lambda \,, \ \lambda \text{ in } [0,\alpha) \,, \\ &\quad\ \ \equiv \alpha \,, \ \lambda \text{ in } [\alpha,\infty) \,, \\ &g(\lambda) \equiv \lambda + \log(1-e^{\lambda-\alpha}) \,, \ \lambda \text{ in } [0,\alpha) \,, \\ &\quad\ \ \equiv -\infty \,, \ \lambda \text{ in } [\alpha,\infty) \,. \end{aligned} \qquad (16)$$

Corollary. Further, in the case of Theorem 2 we have

$$||u(t)-w(t)|| \leq 2\ {}^{t}E^{1-t} \,. \qquad (17)$$

Condition (15) on f is slightly (but it can be shown only slightly) more restrictive than the previous restriction (10) of Theorem 1. Moreover, its formulae are definitely less complicated. For this reason we prefer to work with (15). We will call (15) the S.Q.R. condition on f and the construction of v(t) or w(t) by (2), (3) with a particular choice of such an f will be called an S.Q.R. method. We illustrate that condition in Figure 2.

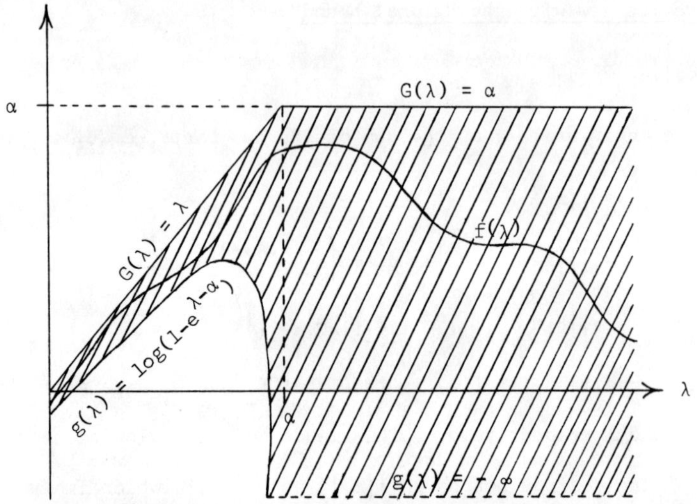

Figure 2. The S.Q.R. condition (15) on f

5. POLYNOMIAL CHOICES FOR f

If the spectral decomposition of A is known then the S.Q.R. construction (2) is fairly easily accomplished for essentially any choice of f satisfying (15). However, in that case it would be better to use the previous method of partial eigenfunction expansion. Our hope for S.Q.R. is that it will lead to methods which are more efficiently computable than previous ones. For this reason then, we are interested in having polynomial choices for f.

We point out first that the order of the polynomial need not be extremely large. One can show that

$$f(\lambda) = \lambda - (\alpha^{-\beta})\lambda^{\beta} \ , \ \beta \text{ any integer} \geq \alpha \ , \tag{18}$$

will satisfy the S.Q.R. condition (15).

I doubt that an f of order greatly less than α can be found. Nevertheless, one can see that our polynomial of (18) is not of optimally low order; by an extension of the usual Chebyschev type analysis it can be shown that an f of optimally low order p* should have a more oscillatory type behaviour, oscillating back and forth p* + 2 times between the curve $G(\lambda)$ above and the curve $g(\lambda)$ below.

6. RATIONAL FINITE DIFFERENCE METHODS FOR THE S.Q.R. CONSTRUCTION

From the computational point of view we should not be satisfied with just having found a sufficiently low order polynomial for f, for approximation of $e^{(1-t)f(A)}$ still presents great computational difficulties since f(A) will have gigantic or unbounded negative spectrum.

Our first approach is to fix a polynomial f satisfying (15) then introduce finite difference methods to approximate the exponential $e^{(1-t)B}$, with B = f(A), to within "permissible accuracy". That is, we divide the t interval into n intervals of length $\Delta t = 1/n$, let $v_j = v(j\Delta t)$, then use a finite difference method to approximate

$$v_j = e^{\Delta t f(A)} v_{j+1}, \qquad \text{(exact)}$$

$$\text{or} \quad \approx (I+\Delta t f(A))^{-1} v_{j+1}, \qquad \text{(backward difference equation)}$$

$$\text{or} \quad \approx \left\{\frac{I + \frac{\Delta t}{2} f(A)}{I - \frac{\Delta t}{2} f(A)}\right\} v_{j+1}, \qquad \text{(trapezoid rule)} \tag{19}$$

$$\text{or} \quad \approx Q(\Delta t f(A)) v_{j+1},$$

where $Q(\chi)$ is some higher order rational approximant to e^{χ}. In this way, we can obtain some fairly satisfactory methods.

It is much more worthwhile however, to note that our approximation has always been of the form

$$v_j = R(A) v_{j+1}, \tag{20}$$

where $R(\lambda)$ is a rational function. Let us suppose that $R(\lambda)$ is exactly equal $e^{\Delta t f(\lambda)}$, where f is not necessarily a polynomial. Now <u>$f(\lambda)$ satisfies</u>

the S.Q.R. condition (15) if and only if

$$e^{\Delta t g(\lambda)} \lesssim e^{\Delta t f(\lambda)} \equiv R(\lambda) \lesssim e^{\Delta t G(\lambda)}$$

or

$$e^{g(\lambda)} \lesssim (R(\lambda))^n \lesssim e^{G(\lambda)} \quad , \ \Delta t = \frac{1}{n} \ . \tag{21}$$

We call (21) the S.Q.R. finite difference condition on the rational function R, and illustrate it in Figure 3.

Thus, instead of first looking for polynomial or rational functions f satisfying (15), we need only look directly for rational functions satisfying (21).

The following gives one explicit construction of an R of "order of accuracy" 2q, where q may be arbitrarily large.

Lemma. (An explicit construction of an R.) Let

$$R = Q(\Delta t P(\lambda)) \ , \text{ where} \qquad \text{(a)}$$

$$P(\lambda) = \lambda - (\alpha/\alpha+1)^{\alpha}(\alpha^{-\beta})\lambda^{\beta} \ , \qquad \text{(b)} \tag{22}$$

where β is any integer $\geq \alpha + 1$ and $Q(\chi)$ is the qth order diagonal Padé approximant to e^{χ}, with q even. Then R satisfies (21) provided

$$(\Delta t)^{2q} \lesssim c \cdot (\varepsilon/E)$$

where c is a certain constant.

The rational methods just presented are still probably far from optimal. It should probably be possible to use rational best approximation theory to find the R of lowest possible order which fits between the two curves in Figure 3. Two colleagues and I are looking into this now.

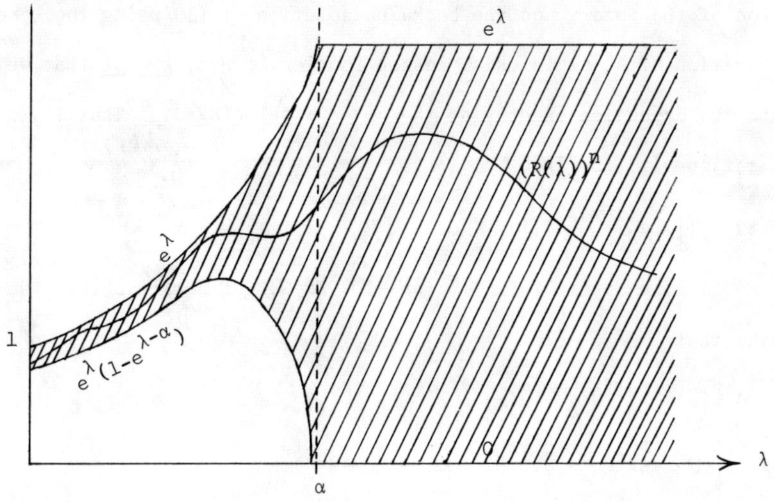

Figure 3. The S.Q.R. finite difference condition (21) on R

In all cases we would recommend that the inversions involved in computing $v_j = R(A)v_{j+1}$ should be carried out one linear complex factor at a time. That is, one should compute the complex roots $1/\xi_i$ and $1/\eta_i$ of the numerator and denominator to great accuracy by a standard polynomial root program, and then carry out the computation in the factored order

$$v_j = k\left(\frac{I-\xi_r A}{I-\eta_r A}\left(\cdots\left(\frac{I-\xi_2 A}{I-\eta_2 A}\left(\frac{I-\xi_1 A}{I-\eta_1 A}v_{j+1}\right)\right.\right.\right. . \tag{23}$$

Thus, one needs only to compute resolvents of A. Alternatively, one might instead prefer to carry out the inversions one real quadratic factor at a time, thereby avoiding complex arithmetic.

7. POSSIBLE COMPUTATIONAL ADVANTAGES OF S.Q.R. OVER PARTIAL EXPANSION AND LEAST SQUARES

The method of partial eigenfunction expansion, when applied to the backward solution of (1) under the hypotheses (6), turns out to be an extreme case of S.Q.R. The method in brief is the following: we carry out the expansion of the data χ and the backward solution of (1) using the spectral representation of A, as is usual in the Fourier method, except that we truncate the expansion exactly at the order $\alpha = \log(E/\varepsilon)$. That is, our approximation v(t) is given by

$$v(t) = \left(\int_0^\alpha e^{(1-t)\lambda} dE_\lambda\right)\chi \tag{24}$$

where $\{E_\lambda\}$ is the spectral family for A. In S.Q.R. notation then, we recognize that

$$v(t) = e^{(1-t)f(A)}\chi = w(t) = e^{-tA}(e^{f(A)}\chi) \; , \tag{25}$$

where f is the extreme S.Q.R. choice

$$f(\lambda) \equiv \lambda \text{ on } [0,\alpha] \; , \equiv -\infty \text{ on } (\alpha,\infty) \; . \tag{26}$$

The method of least squares may be summarized as follows: our approximation w(t) is the exact solution of (1) forward, $w(t) \equiv e^{-tA}\xi$, from an initial vector obtained by minimizing $||e^{-A}\xi-\chi||^2 + (\varepsilon/E)^2||\xi||^2$ among all ξ in H. That is, ξ is the solution of the least squares canonical equations

$$(e^{-2A}+(\varepsilon/E)^2 I)\xi = e^{-A}\chi \; . \tag{27}$$

The error, as in (17) is bounded by $\sqrt{2}\varepsilon^t E^{1-t}$.

When the spectral decomposition of A is readily available, such as is often the case in many problems of partial differential equations for which an explicit separation of variables solution is possible, then the partial expansion method is definitely to be recommended. However, except in very special circumstances, constant coefficients and simple geometries for example, the eigenfunctions and eigenvalues of a partial differential operator A just will not be readily available.

Let us consider the typical example $u' = L_h u - Au$ mentioned in the introduction. (We assume that the original p.d.e. problem $u' = Lu$ has already been discretized in χ sufficiently accurately for our purposes.)

In this case A will be a very large matrix (a 50×50 finite difference grid in a 2-space dimension problem would make A 2500×2500 for example), but with the fortunate feature that it is sparse, that is, most of its entries are zero.

The least squares method in this high dimensionality situation has the almost insurmountable difficulty that it fails to make use of the sparseness of A. The matrix $e^{-2A} + (\varepsilon/E)^2 I$ involved, still 2500×2500 in size, will in general have no zero elements.

The rational finite difference S.Q.R. methods, however, proceed backward from step to step by inverting one complex linear factor $(I-\eta_i A)$ at a time (or one real quadratic factor $(I-\eta_i A)(I-\bar{\eta}_i A)$ at a time). Thus, S.Q.R. takes full advantage of the sparseness of A, thereby making feasible problems of a size which would be impossible by the previous methods.

8. EXTENSIONS

Our results extend almost immediately to normal A with non-negative real parts. It seems hopeful (but this has not yet been done) that S.Q.R. type methods could be applied to A which are "sectorial operators", or "generators of holomorphic semigroups". These include most elliptic partial differential operators, of arbitrary order. For such A one can quite easily derive backward stability results [9] similar to those given by logarithmic convexity for self-adjoint A.

Preliminary considerations seems to indicate, however, that efficient S.Q.R. type methods should not be possible for time dependent A(t). In that case one would probably have to turn to the more general least squares methods to get optimal error results.

REFERENCES

[1] Agmon, S., and Nirenberg, L., Properties of solutions of ordinary differential equations in Banach space, Comm. Pure Appl. Math., 16 (1963), pp. 121-239. (see p. 136)

[2] Backus, G., Inference from inadequate and inaccurate data, I, Proc. Nat. Acad. Sci. U.S.A., 65 (1970), pp. 1-7.

[3] Douglas, J., A numerical method for analytic continuation, Boundary Problems in Differential Equations, Univ. of Wisconsin Press, Madison, 1960, pp. 179-189.

[4] John, F., Continuous dependence on data for solutions of partial differential equations with a prescribed bound, Comm. Pure Appl. Math., 13 (1960), pp. 551-585.

[5] Lattes, R., and Lions, J., Méthodé de Quasi-Réversibilité et Applications, Dunod, Paris, 1967.

[6] Miller, K., Three circle theorems in partial differential equations and applications to improperly posed problems, Arch. Rational Mech. Anal., 16 (1964), pp. 126-154.

[7] Miller, K., Least squares methods for ill-posed problems with a prescribed bound, SIAM J. Math. Anal., 1 (1970), pp. 52-73.

[8] Miller, K., Stabilized version of the method of quasi-réversibilité, (to appear).

[9] Miller, K., Logarithmic convexity results for holomorphic semigroups, (in preparation).

[10] Miller, K., and Viano, G., On the necessity of nearly-best-p methods for the analytic continuation of scattering data, (to appear).

[11] Pucci, C., Studio col metodo delle differenze di un problema di Cauchy relativo ad equazioni a derivate parziali del secondo ordine di tipo parabolico, Ann. della Scuola Norm. Sup. di Pisa, Serie III, Vol. VII, Fasc. III-IV (1953), pp. 205-215.

[12] Pucci, C., Sui problemi di Cauchy non "ben posti", Atti. Accad. Naz. Lincei Rend. Al. Sci. Fis. Mat. Natur. (8), 18 (1955), pp. 473-477.

Department of Mathematics

University of California

Berkeley, California

Lecture Notes in Mathematics

Comprehensive leaflet on request

Vol. 146: A. B. Altman and S. Kleiman, Introduction to Grothendieck Duality Theory. II, 192 pages. 1970. DM 18,–

Vol. 147: D. E. Dobbs, Cech Cohomological Dimensions for Commutative Rings. VI, 176 pages. 1970. DM 16,–

Vol. 148: R. Azencott, Espaces de Poisson des Groupes Localement Compacts. IX, 141 pages. 1970. DM 16,–

Vol. 149: R. G. Swan and E. G. Evans, K-Theory of Finite Groups and Orders. IV, 237 pages. 1970. DM 20,–

Vol. 150: Heyer, Dualität lokalkompakter Gruppen. XIII, 372 Seiten. 1970. DM 20,–

Vol. 151: M. Demazure et A. Grothendieck, Schémas en Groupes I. (SGA 3). XV, 562 pages. 1970. DM 24,–

Vol. 152: M. Demazure et A. Grothendieck, Schémas en Groupes II. (SGA 3). IX, 654 pages. 1970. DM 24,–

Vol. 153: M. Demazure et A. Grothendieck, Schémas en Groupes III. (SGA 3). VIII, 529 pages. 1970. DM 24,–

Vol. 154: A. Lascoux et M. Berger, Variétés Kähleriennes Compactes. VII, 83 pages. 1970. DM 16,–

Vol. 155: Several Complex Variables I, Maryland 1970. Edited by J. Horváth. IV, 214 pages. 1970. DM 18,–

Vol. 156: R. Hartshorne, Ample Subvarieties of Algebraic Varieties. XIV, 256 pages. 1970. DM 20,–

Vol. 157: T. tom Dieck, K. H. Kamps und D. Puppe, Homotopietheorie. VI, 265 Seiten. 1970. DM 20,–

Vol. 158: T. G. Ostrom, Finite Translation Planes. IV. 112 pages 1970. DM 16,–

Vol. 159: R. Ansorge und R. Hass. Konvergenz von Differenzenverfahren für lineare und nichtlineare Anfangswertaufgaben. VIII, 145 Seiten. 1970. DM 16,–

Vol. 160: L. Sucheston, Constributions to Ergodic Theory and Probability. VII, 277 pages. 1970. DM 20,–

Vol. 161: J. Stasheff, H-Spaces from a Homotopy Point of View. VI, 95 pages. 1970. DM 16,–

Vol. 162: Harish-Chandra and van Dijk, Harmonic Analysis on Reductive p-adic Groups. IV, 125 pages. 1970. DM 16,–

Vol. 163: P. Deligne, Equations Différentielles à Points Singuliers Reguliers. III, 133 pages. 1970. DM 16,–

Vol. 164: J. P. Ferrier, Seminaire sur les Algebres Complètes. II, 69 pages. 1970. DM 16,–

Vol. 165: J. M. Cohen, Stable Homotopy. V, 194 pages. 1970. DM 16.–

Vol. 166: A. J. Silberger, PGL_2 over the p-adics: its Representations, Spherical Functions, and Fourier Analysis. VII, 202 pages. 1970. DM 18,–

Vol. 167: Lavrentiev, Romanov and Vasiliev, Multidimensional Inverse Problems for Differential Equations. V, 59 pages. 1970. DM 16,–

Vol. 168: F. P. Peterson, The Steenrod Algebra and its Applications: A conference to Celebrate N. E. Steenrod's Sixtieth Birthday. VII, 317 pages. 1970. DM 22,–

Vol. 169: M. Raynaud, Anneaux Locaux Henséliens. V, 129 pages. 1970. DM 16,–

Vol. 170: Lectures in Modern Analysis and Applications III. Edited by C. T. Taam. VI, 213 pages. 1970. DM 18,–

Vol. 171: Set-Valued Mappings, Selections and Topological Properties of 2^X. Edited by W. M. Fleischman. X, 110 pages. 1970. DM 16,–

Vol. 172: Y.-T. Siu and G. Trautmann, Gap-Sheaves and Extension of Coherent Analytic Subsheaves. V, 172 pages. 1971. DM 16,–

Vol. 173: J. N. Mordeson and B. Vinograde, Structure of Arbitrary Purely Inseparable Extension Fields. IV, 138 pages. 1970. DM 16,–

Vol. 174: B. Iversen, Linear Determinants with Applications to the Picard Scheme of a Family of Algebraic Curves. VI, 69 pages. 1970. DM 16,–

Vol. 175: M. Brelot, On Topologies and Boundaries in Potential Theory. VI, 176 pages. 1971. DM 18,–

Vol. 176: H. Popp, Fundamentalgruppen algebraischer Mannigfaltigkeiten. IV, 154 Seiten. 1970. DM 16,–

Vol. 177: J. Lambek, Torsion Theories, Additive Semantics and Rings of Quotients. VI, 94 pages. 1971. DM 16,–

Vol. 178: Th. Bröcker und T. tom Dieck, Kobordismentheorie. XVI, 191 Seiten. 1970. DM 18,–

Vol. 179: Seminaire Bourbaki – vol. 1968/69. Exposés 347-363. IV. 295 pages. 1971. DM 22,–

Vol. 180: Séminaire Bourbaki – vol. 1969/70. Exposés 364-381. IV, 310 pages. 1971. DM 22,–

Vol. 181: F. DeMeyer and E. Ingraham, Separable Algebras over Commutative Rings. V, 157 pages. 1971. DM 16.–

Vol. 182: L. D. Baumert. Cyclic Difference Sets. VI, 166 pages. 1971. DM 16,–

Vol. 183: Analytic Theory of Differential Equations. Edited by P. F. Hsieh and A. W. J. Stoddart. VI, 225 pages. 1971. DM 20,–

Vol. 184: Symposium on Several Complex Variables, Park City, Utah, 1970. Edited by R. M. Brooks. V, 234 pages. 1971. DM 20,–

Vol. 185: Several Complex Variables II, Maryland 1970. Edited by J. Horváth. III, 287 pages. 1971. DM 24,–

Vol. 186: Recent Trends in Graph Theory. Edited by M. Capobianco/ J. B. Frechen/M. Krolik. VI, 219 pages. 1971. DM 18.–

Vol. 187: H. S. Shapiro, Topics in Approximation Theory. VIII, 275 pages. 1971. DM 22,–

Vol. 188: Symposium on Semantics of Algorithmic Languages. Edited by E. Engeler. VI, 372 pages. 1971. DM 26,–

Vol. 189: A. Weil, Dirichlet Series and Automorphic Forms. V, 164 pages. 1971. DM 16,–

Vol. 190: Martingales. A Report on a Meeting at Oberwolfach, May 17-23, 1970. Edited by H. Dinges. V, 75 pages. 1971. DM 16,–

Vol. 191: Séminaire de Probabilités V. Edited by P. A. Meyer. IV, 372 pages. 1971. DM 26,–

Vol. 192: Proceedings of Liverpool Singularities – Symposium I. Edited by C. T. C. Wall. V, 319 pages. 1971. DM 24,–

Vol. 193: Symposium on the Theory of Numerical Analysis. Edited by J. Ll. Morris. VI, 152 pages. 1971. DM 16,–

Vol. 194: M. Berger, P. Gauduchon et E. Mazet. Le Spectre d'une Variété Riemannienne. VII, 251 pages. 1971. DM 22,–

Vol. 195: Reports of the Midwest Category Seminar V. Edited by J.W. Gray and S. Mac Lane.III, 255 pages. 1971. DM 22,–

Vol. 196: H-spaces – Neuchâtel (Suisse)- Août 1970. Edited by F. Sigrist, V, 156 pages. 1971. DM 16,–

Vol. 197: Manifolds – Amsterdam 1970. Edited by N. H. Kuiper. V, 231 pages. 1971. DM 20,–

Vol. 198: M. Hervé, Analytic and Plurisubharmonic Functions in Finite and Infinite Dimensional Spaces. VI, 90 pages. 1971. DM 16.–

Vol. 199: Ch. J. Mozzochi, On the Pointwise Convergence of Fourier Series. VII, 87 pages. 1971. DM 16,–

Vol. 200: U. Neri, Singular Integrals. VII, 272 pages. 1971. DM 22,–

Vol. 201: J. H. van Lint, Coding Theory. VII, 136 pages. 1971. DM 16,–

Vol. 202: J. Benedetto, Harmonic Analysis on Totally Disconnected Sets. VIII, 261 pages. 1971. DM 22,–

Vol. 203: D. Knutson, Algebraic Spaces. VI, 261 pages. 1971. DM 22,–

Vol. 204: A. Zygmund, Intégrales Singulières. IV, 53 pages. 1971. DM 16,–

Vol. 205: Séminaire Pierre Lelong (Analyse) Année 1970. VI, 243 pages. 1971. DM 20,–

Vol. 206: Symposium on Differential Equations and Dynamical Systems. Edited by D. Chillingworth. XI, 173 pages. 1971. DM 16,–

Vol. 207: L. Bernstein, The Jacobi-Perron Algorithm – Its Theory and Application. IV, 161 pages. 1971. DM 16,–

Vol. 208: A. Grothendieck and J. P. Murre, The Tame Fundamental Group of a Formal Neighbourhood of a Divisor with Normal Crossings on a Scheme. VIII, 133 pages. 1971. DM 16,–

Vol. 209: Proceedings of Liverpool Singularities Symposium II. Edited by C. T. C. Wall. V, 280 pages. 1971. DM 22,–

Vol. 210: M. Eichler, Projective Varieties and Modular Forms. III, 118 pages. 1971. DM 16,–

Vol. 211: Théorie des Matroïdes. Edité par C. P. Bruter. III, 108 pages. 1971. DM 16,–

Please turn over

Vol. 212: B. Scarpellini, Proof Theory and Intuitionistic Systems. VII, 291 pages. 1971. DM 24,–

Vol. 213: H. Hogbe-Nlend, Théorie des Bornologies et Applications. V, 168 pages. 1971. DM 18,–

Vol. 214: M. Smorodinsky, Ergodic Theory, Entropy. V, 64 pages. 1971. DM 16,–

Vol. 215: P. Antonelli, D. Burghelea and P. J. Kahn, The Concordance-Homotopy Groups of Geometric Automorphism Groups. X, 140 pages. 1971. DM 16,–

Vol. 216: H. Maaß, Siegel's Modular Forms and Dirichlet Series. VII, 328 pages. 1971. DM 20,–

Vol. 217: T. J. Jech, Lectures in Set Theory with Particular Emphasis on the Method of Forcing. V, 137 pages. 1971. DM 16,–

Vol. 218: C. P. Schnorr, Zufälligkeit und Wahrscheinlichkeit. IV, 212 Seiten 1971. DM 20,–

Vol. 219: N. L. Alling and N. Greenleaf, Foundations of the Theory of Klein Surfaces. IX, 117 pages. 1971. DM 16,–

Vol. 220: W. A. Coppel, Disconjugacy. V, 148 pages. 1971. DM 16,–

Vol. 221: P. Gabriel und F. Ulmer, Lokal präsentierbare Kategorien. V, 200 Seiten. 1971. DM 18,–

Vol. 222: C. Meghea, Compactification des Espaces Harmoniques. III, 108 pages. 1971. DM 16,–

Vol. 223: U. Felgner, Models of ZF-Set Theory. VI, 173 pages. 1971. DM 16,–

Vol. 224: Revètements Etales et Groupe Fondamental. (SGA 1). Dirigé par A. Grothendieck XXII, 447 pages. 1971. DM 30,–

Vol. 225: Théorie des Intersections et Théorème de Riemann-Roch. (SGA 6). Dirigé par P. Berthelot, A. Grothendieck et L. Illusie. XII, 700 pages. 1971. DM 40,–

Vol. 226: Seminar on Potential Theory, II. Edited by H. Bauer. IV, 170 pages. 1971. DM 18,–

Vol. 227: H. L. Montgomery, Topics in Multiplicative Number Theory. IX, 178 pages. 1971. DM 18,–

Vol. 228: Conference on Applications of Numerical Analysis. Edited by J. Ll. Morris. X, 358 pages. 1971. DM 26,–

Vol. 229: J. Väisälä, Lectures on n-Dimensional Quasiconformal Mappings. XIV, 144 pages. 1971. DM 16,–

Vol. 230: L. Waelbroeck, Topological Vector Spaces and Algebras. VII, 158 pages. 1971. DM 16,–

Vol. 231: H. Reiter, L^1-Algebras and Segal Algebras. XI, 113 pages. 1971. DM 16,–

Vol. 232: T. H. Ganelius, Tauberian Remainder Theorems. VI, 75 pages. 1971. DM 16,–

Vol. 233: C. P. Tsokos and W. J. Padgett. Random Integral Equations with Applications to Stochastic Systems. VII, 174 pages. 1971. DM 18,–

Vol. 234: A. Andreotti and W. Stoll. Analytic and Algebraic Dependence of Meromorphic Functions. III, 390 pages. 1971. DM 26,–

Vol. 235: Global Differentiable Dynamics. Edited by O. Hájek, A. J. Lohwater, and R. McCann. X, 140 pages. 1971. DM 16,–

Vol. 236: M. Barr, P. A. Grillet, and D. H. van Osdol. Exact Categories and Categories of Sheaves. VII, 239 pages. 1971, DM 20,–

Vol. 237: B. Stenström. Rings and Modules of Quotients. VII, 136 pages. 1971. DM 16,–

Vol. 238: Der kanonische Modul eines Cohen-Macaulay-Rings. Herausgegeben von Jürgen Herzog und Ernst Kunz. VI, 103 Seiten. 1971. DM 16,–

Vol. 239: L. Illusie, Complexe Cotangent et Déformations I. XV, 355 pages. 1971. DM 26,–

Vol. 240: A. Kerber, Representations of Permutation Groups I. VII, 192 pages. 1971. DM 18,–

Vol. 241: S. Kaneyuki, Homogeneous Bounded Domains and Siegel Domains. V, 89 pages. 1971. DM 16,–

Vol. 242: R. R. Coifman et G. Weiss, Analyse Harmonique Non-Commutative sur Certains Espaces. V, 160 pages. 1971. DM 16,–

Vol. 243: Japan-United States Seminar on Ordinary Differential and Functional Equations. Edited by M. Urabe. VIII, 332 pages. 1971. DM 26,–

Vol. 244: Séminaire Bourbaki – vol. 1970/71. Exposés 382–399. IV, 356 pages. 1971. DM 26,–

Vol. 245: D. E. Cohen, Groups of Cohomological Dimension One. V, 99 pages. 1972. DM 16,–

Vol. 246: Lectures on Rings and Modules. Tulane University Ring and Operator Theory Year, 1970–1971. Volume I. X, 661 pages. 1972. DM 40,–

Vol. 247: Lectures on Operator Algebras. Tulane University Ring and Operator Theory Year, 1970–1971. Volume II. XI, 786 pages. 1972. DM 40,–

Vol. 248: Lectures on the Applications of Sheaves to Ring Theory. Tulane University Ring and Operator Theory Year, 1970–1971. Volume III. VIII, 315 pages. 1971. DM 26,–

Vol. 249: Symposium on Algebraic Topology. Edited by P. J. Hilton. VII, 111 pages. 1971. DM 16,–

Vol. 250: B. Jónsson, Topics in Universal Algebra. VI, 220 pages. 1972. DM 20,–

Vol. 251: The Theory of Arithmetic Functions. Edited by A. A. Gioia and D. L. Goldsmith VI, 287 pages. 1972. DM 24,–

Vol. 252: D. A. Stone, Stratified Polyhedra. IX, 193 pages. 1972. DM 18,–

Vol. 253: V. Komkov, Optimal Control Theory for the Damping of Vibrations of Simple Elastic Systems. V, 240 pages. 1972. DM 20,–

Vol. 254: C. U. Jensen, Les Foncteurs Dérivés de $\varprojlim$ et leurs Applications en Théorie des Modules. V, 103 pages. 1972. DM 16,–

Vol. 255: Conference in Mathematical Logic – London '70. Edited by W. Hodges. VIII, 351 pages. 1972. DM 26,–

Vol. 256: C. A. Berenstein and M. A. Dostal, Analytically Uniform Spaces and their Applications to Convolution Equations. VII, 130 pages. 1972. DM 16,–

Vol. 257: R. B. Holmes, A Course on Optimization and Best Approximation. VIII, 233 pages. 1972. DM 20,–

Vol. 258: Séminaire de Probabilités VI. Edited by P. A. Meyer. VI, 253 pages. 1972. DM 22,–

Vol. 259: N. Moulis, Structures de Fredholm sur les Variétés Hilbertiennes. V, 123 pages. 1972. DM 16,–

Vol. 260: R. Godement and H. Jacquet, Zeta Functions of Simple Algebras. IX, 188 pages. 1972. DM 18,–

Vol. 261: A. Guichardet, Symmetric Hilbert Spaces and Related Topics. V, 197 pages. 1972. DM 18,–

Vol. 262: H. G. Zimmer, Computational Problems, Methods, and Results in Algebraic Number Theory. V, 103 pages. 1972. DM 16,–

Vol. 263: T. Parthasarathy, Selection Theorems and their Applications. VII, 101 pages. 1972. DM 16,–

Vol. 264: W. Messing, The Crystals Associated to Barsotti-Tate Groups: with Applications to Abelian Schemes. III, 190 pages. 1972. DM 18,–

Vol. 265: N. Saavedra Rivano, Catégories Tannakiennes. II, 418 pages. 1972. DM 26,–

Vol. 266: Conference on Harmonic Analysis. Edited by D. Gulick and R. L. Lipsman. VI, 323 pages. 1972. DM 24,–

Vol. 267: Numerische Lösung nichtlinearer partieller Differential- und Integro-Differentialgleichungen. Herausgegeben von R. Ansorge und W. Törnig, VI, 339 Seiten. 1972. DM 26,–

Vol. 268: C. G. Simader, On Dirichlet's Boundary Value Problem. IV, 238 pages. 1972. DM 20,–

Vol. 269: Théorie des Topos et Cohomologie Etale des Schémas. (SGA 4). Dirigé par M. Artin, A. Grothendieck et J. L. Verdier. XIX, 525 pages. 1972. DM 50,–

Vol. 270: Théorie des Topos et Cohomologie Etle des Schémas. Tome 2. (SGA 4). Dirige par M. Artin, A. Grothendieck et J. L. Verdier. V, 418 pages. 1972. DM 50,–

Vol. 271: J. P. May, The Geometry of Iterated Loop Spaces. IX, 175 pages. 1972. DM 18,–

Vol. 272: K. R. Parthasarathy and K. Schmidt, Positive Definite Kernels, Continuous Tensor Products, and Central Limit Theorems of Probability Theory. VI, 107 pages. 1972. DM 16,–

Vol. 273: U. Seip, Kompakt erzeugte Vektorräume und Analysis. IX, 119 Seiten. 1972. DM 16,–

Vol. 274: Toposes, Algebraic Geometry and Logic. Edited by. F. W. Lawvere. VI, 189 pages. 1972. DM 18,–

Vol. 275: Séminaire Pierre Lelong (Analyse) Année 1970–1971. VI, 181 pages. 1972. DM 18,–

Vol. 276: A. Borel, Représentations de Groupes Localement Compacts. V, 98 pages. 1972. DM 16,–

Vol. 277: Séminaire Banach. Edité par C. Houzel. VII, 229 pages. 1972. DM 20,–